Design Automation for Differential MOS Current-Mode Logic Circuits

Stéphane Badel • Can Baltaci • Alessandro Cevrero
Yusuf Leblebici

Design Automation for Differential MOS Current-Mode Logic Circuits

Springer

Stéphane Badel
École Polytechnique Fédérale de Lausanne
Lausanne, Switzerland

Can Baltaci
École Polytechnique Fédérale de Lausanne
Lausanne, Switzerland

Alessandro Cevrero
IBM Research – Zurich
Rüschlikon, Switzerland

Yusuf Leblebici
École Polytechnique Fédérale de Lausanne
Lausanne, Switzerland

ISBN 978-3-030-08219-2 ISBN 978-3-319-91307-0 (eBook)
https://doi.org/10.1007/978-3-319-91307-0

Printed on acid-free paper

This Springer imprint is published by the registered company Springer International Publishing AG part of Springer Nature.
The registered company address is: Gewerbestrasse 11, 6330 Cham, Switzerland

Preface

Our detailed research work on the design and optimization of high-performance MOS current mode logic (MCML) circuits at the Microelectronic Systems Laboratory (LSM) of EPFL started more than a decade ago. In the beginning, our main motivation was the reduction of power supply noise and substrate noise generated by high-speed logic units that have to operate in very close proximity to sensitive analog building blocks. While the fundamental concepts used in the design of MCML circuits were fairly well understood, relatively little work was available at that time to guide systematic analysis and especially design automation of such circuits. Our early research in this domain has led to the development of differential logic cell optimization techniques under arbitrary load conditions, as well as fully differential logic synthesis, and placement-and-routing (P&R) strategies that enable straightforward design automation of logic functions based on conventional hardware description languages such as VHDL and Verilog. Such logic units distinguish themselves with their capability of operating at multi-GHz frequencies while producing extremely low levels of supply noise. Nowadays, MCML-based circuit solutions are commonly used in various applications where high-performance operation is the primary objective.

In addition to high-speed operation, the fully differential nature of the MCML circuit style lends itself to implementation of logic blocks in which the power supply signature associated with the logic operations can be effectively suppressed. This property results in highly efficient implementation of various cryptographic functions with a remarkable immunity to differential power analysis (DPA) attacks. The fully differential current-mode operation principle of MCML circuits has also paved the way for the development of a completely new class of ultralow-power logic circuits called sub-threshold source-coupled logic (ST-SCL) which can achieve impressive energy efficiency operating with very low tail current levels (down to a few pA) while producing several hundreds of mV output voltage swing—a feature that is simply not possible in conventional CMOS logic circuits operating in sub-threshold regime. Our extensive work in this particular direction has already been published in the form of a separate volume from Springer, entitled *Extreme Low-Power Mixed Signal IC Design* coauthored by A. Tajalli and Y. Leblebici (ISBN 978-1-4419-6477-9).

This volume covers systematic, in-depth analysis of MCML circuits in Part I (Chaps. 2 and 3), followed by the principles of design automation for MCML in Part II (Chaps. 4 and 5), addressing fully differential logic synthesis, standard cell design, pin assignment, and placement-and-routing strategies. The last four chapters (Part III) of the book are dedicated to specific design examples for high-speed design as well as cryptographic circuit applications, such as the DPA-resistant implementations of Grain-128 stream cipher and AES engines. The topics covered in this book would be beneficial to graduate students specializing in high-speed circuit design, as well as engineering professionals designing systems for high performance and DPA immunity.

The authors are truly indebted to many individuals who have contributed to this work. Our graduate students, as well as our colleagues, have consistently helped us with their generous assistance along the way. In particular, we acknowledge the valuable support provided over the years by Dr. Ilhan Hatirnaz, Dr. Francesco Regazzoni, Dr. Armin Tajalli, Ms. Tugba Demirci, and Mr. Michael Schwander. This work would not have been possible without their contributions.

Lausanne, Switzerland — Stéphane Badel
Lausanne, Switzerland — Can Baltaci
Rüschlikon, Switzerland — Alessandro Cevrero
Lausanne, Switzerland — Yusuf Leblebici
14 May 2018

Contents

Chapter 1
Introduction

Over the past decades, integrated circuits have evolved from circuits combining thousands of transistors to multi-billion devices in today's advanced technologies. The continuous scaling of device dimensions in VLSI is enabling the integration of complete systems on a single die, which may include a combination of RF transceivers, analog processing, A/D and D/A conversion as well as complex digital functions and memory on a single chip.

Combining all these elements on a single chip has many advantages, including reduced cost, higher speed, and lower overall power dissipation. It does not come, however, without its very own problems, not the least of which is the increase in noise coupled from the digital functions to the analog parts.

1.1 Noise in Integrated Circuits

When sensitive analog parts are combined with complex digital blocks operating at very high switching frequencies, the noise generated by the digital parts is inevitably transmitted to the analog blocks, predominantly through the common substrate, resulting in a reduction of the dynamic range, or reduction of the accuracy of the analog circuits.

Noise in digital CMOS circuits is mainly generated by the rapid voltage variations caused by the switching of logic states, and the related charge-up / charge-down currents. In a conventional CMOS logic gate, the rapid change of voltage in internal nodes is coupled to the substrate through junction or wiring capacitances, causing charges to be injected into the substrate. Eventually, these substrate currents cause voltage drops that can perturb analog circuits through capacitive coupling and through variation of the threshold voltage due to body effect [8, 11].

Additionally, the high instantaneous currents needed to rapidly charge or discharge parasitic capacitances add up a large current spikes in the supply and ground

S. Badel et al., *Design Automation for Differential MOS Current-Mode Logic Circuits*, https://doi.org/10.1007/978-3-319-91307-0_1

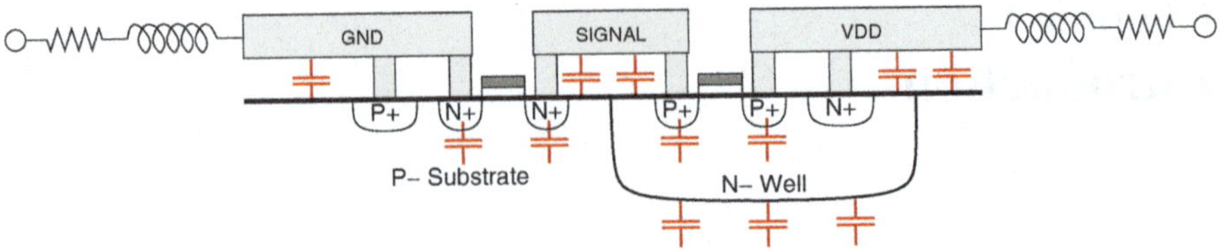

Fig. 1.1 Schematic cross-section of a typical N-well CMOS process illustrating the different mechanisms of noise coupling between power and signal nets and the substrate

distribution networks, a phenomenon known an simultaneous switching noise (SSN). These current spikes cause voltage noise primarily through the inductance of off-chip bond-wires and on-chip power-supply rails. Ground supply networks are usually directly connected to the substrate, resulting in a direct coupling of the noise, and power networks are typically connected to very large N-well areas resulting in a consequently very large parasitic coupling capacitance to the substrate. Therefore, power and ground distribution networks are very noisy in CMOS circuits, and at the same time ideal mediums for the noise coupling to the substrate. Signal nets can also couple to the substrate, through diffusion and wiring capacitances, and signals with high energy and switching activity are thus critical from a noise perspective. This is the case especially for clock networks, which are the most active signals and dissipate large amounts of power (Fig. 1.1).

1.2 Low-Noise CMOS Logic Families

Two effective techniques to reduce the noise generation in digital circuits are the reduction of the voltage swings, and the cancellation of transient currents during switching events. In the past few years, several new logic families have been proposed, that generate less noise than classical CMOS logic, and are thus suitable for integration in mixed-mode environment as a replacement or a complement of CMOS logic.

These new logic families can broadly be categorized into two classes:

- single-ended families, including Current Steering Logic (CSL) [9] and Current-Balanced Logic (CBL) [1], which are based on regular CMOS operation with the addition of a circuitry to limit or cancel the current transients,
- differential families, including Complementary Current Balanced Logic (C-CBL) [2], Folded Source-Coupled Logic (FSCL) [4], and MOS Current-Mode Logic (MCML) [12], where each transition is canceled by an equal and opposite complementary signal.

Experimental studies have shown that single-ended families achieve only marginal improvement over regular CMOS in terms of noise [3]. While differential logic families are the most promising candidates that offer improved noise reduction

[2], traditional automation tools and design flows fail to accommodate many aspects associated with their differential nature. For this reason, large-scale implementation of digital circuits with low-noise differential logic families remains a difficult task, and few results have been reported yet.

Even for very specific targets such as the construction and routing of a fully differential clock distribution network, most of the design tasks have to be carried out manually—which inevitably limits the usability of differential techniques.

1.3 MOS Current-Mode Logic

MOS Current-Mode Logic (MCML) has been introduced in [12] as a new design style for high-speed logic circuit. The operation of MCML circuits is based on the principle of re-directing (or switching) the current of a constant current source through a fully differential network of input transistors, and utilizing the reduced-swing voltage drop on a pair of complementary load devices as the output. Therefore, MCML logic style simultaneously offers reduced voltage swings and differential operation, two key characteristics in reducing the generation of switching noise. In addition, MCML allows high-speed operation, and dissipates a constant power independently of the switching frequency.

Due to these advantageous characteristics, MCML gates have been implemented in various demanding applications such as high speed ring oscillators, frequency dividers, phase detectors, etc. [5–7, 10]. However, until now the design style has remained largely case-specific, where transistor sizing and biasing are chosen to satisfy the particular constraints of a demanding design specification, and standardization of components is not considered in a broader context.

1.4 Organization of the Book

This book addresses the practical aspects and issues related to the implementation of MCML-based logic circuits with a standard-cell methodology. The first part concentrates on the analysis and design of MCML circuits at the transistor-level. The second part focuses on higher-level aspects, including the design of standard-cell libraries and the design automation. The third part presents practical design examples with emphasis on low-noise and high-speed operation.

References

1. M.M. Albuquerque, E.F.M. Silva, Current-balanced logic for mixed-signal IC's. IEEE Int. Symp. Circuits Syst. **1**, 274–277 (1999)

2. E.F.M. Albuquerque, M.M. Silva, Evaluation of substrate noise in CMOS and low-noise logic cells, in *IEEE International Symposium on Circuits and Systems (ISCAS)* (2001)
3. E.F.M. Albuquerque, M.M. Silva, A comparison by simulation and by measurement of the substrate noise generated by CMOS, CSL and CBL digital circuits. IEEE Trans. Circuits Syst. **52**, 734–741 (2005)
4. D.J. Allstot, S.-H. Chee, S. Kiaei, M. Shrivastawa, Folded source-coupled logic vs. CMOS static logic for low-noise mixed-signal IC's. IEEE Trans. Circuits Syst. **40**, 553–563 (1993)
5. H.T. Bui, Y. Savaria, 10 GHz PLL using active shunt-peaked MCML gates and improved frequency acquisition XOR phase detector in 0.18μm CMOS, in *IEEE International Workshop on System-on-Chip for Real-Time Applications* (2004)
6. H.T. Bui, Y. Savaria, Shunt-peaking in MCML gates and its application in the design of a 20 Gb/s half-rate phase detector, in *IEEE International Symposium on Circuits and Systems (ISCAS)* (2004)
7. M.P. Houlgate, D.J. Olszewski, K. Abdelhalim, L. MacHeachern, Adaptable MOS current mode logic for use in a multiband RF prescaler, in *IEEE International Symposium on Circuits and Systems (ISCAS)* (2004)
8. S. Kiaei, D. Allstot, K. Hansen, N.K. Verghese, Noise considerations for mixed-signal RF IC transceivers. Wirel. Netw. **4**, 41–53 (1998)
9. H.-T. Ng, D.J. Allstot, CMOS current steering logic for low-voltage mixed-signal integrated circuits. IEEE Trans. Very Large Scale Integr. Syst. **5**, 301–308 (1997)
10. A. Tanabe, M. Umetani, I. Fujiwara, T. Ogura, K. Kataoka, M. Okihara, H. Sakuraba, T. Endoh, F. Masuoka, 0.18-μm CMOS 10-Gb/s multiplexer/demultiplexer ICs using current mode logic with tolerance to threshold voltage fluctuation. IEEE J. Solid State Circuits **36**, 988–996 (2001)
11. M. van Heijningen, J. Compiet, P. Wambacq, S. Donnay, M.G.E. Engels, and I. Bolsens, Analysis and experimental verification of digital substrate noise generation for epi-type substrates. IEEE J. Solid State Circuits **35**, 1002–1008 (2000)
12. M. Yamashina, H. Yamada, An MOS current mode logic (MCML) circuit for low-power sub-GHz processors, in *IEICE Transactions*, E75-C, October 1992

Part I
Analysis and Design of MOS Current-Mode Logic Circuits

Chapter 2
Analysis of MOS Current-Mode Logic Circuits

2.1 The EKV MOSFET Transistor Model

Throughout this chapter, we use a simple version of the EKV MOSFET transistor model [8]. The EKV model is a fully analytical, charge-based model and provides a simple yet accurate model for hand calculations, and a transregional modeling approach where the MOS transistor is first modeled in asymptotic modes, i.e. weak and strong inversion, and an analytical transregional expression is derived through a continuous interpolation function.

2.1.1 *Strong Inversion Regime*

In strong inversion, the drain current in the EKV model (without accounting for the body effect) is given in linear regime by

$$I_D = n \cdot \beta \cdot \left[\frac{V_G - V_T}{n} - \frac{V_D + V_S}{2} \right] \cdot (V_D - V_S) \tag{2.1}$$

where V_G, V_S, and V_D are the gate, source, and drain voltages, respectively, V_T is the threshold voltage, n is the subthreshold slope parameter, and β is the current factor defined as

$$\beta = \mu \cdot \frac{\epsilon_{ox}}{t_{ox}} \cdot \frac{W}{L} \tag{2.2}$$

with μ the carrier mobility, ϵ_{ox} and t_{ox} the dielectric constant and thickness of the gate dielectric, and W and L the width and length of the gate. In saturation regime, when $n \cdot V_D > V_G - V_T$, the drain current has the expression

S. Badel et al., *Design Automation for Differential MOS Current-Mode Logic Circuits*, https://doi.org/10.1007/978-3-319-91307-0_2

$$I_D = \frac{1}{2} \cdot n \cdot \beta \left(\frac{V_G - V_T}{n} - V_S \right)^2 \tag{2.3}$$

The transconductance in strong inversion and saturation regime is given by

$$g_m = \frac{\partial I_D}{\partial V_G} = \sqrt{\frac{2 \cdot \beta \cdot I_D}{n}} \tag{2.4}$$

2.1.2 *Weak Inversion Regime*

In weak inversion, the drain current in the EKV model (without accounting for the body effect) is given by

$$I_D = I_S \cdot \mathrm{e}^{\frac{V_G - V_T}{n \cdot U_T}} \cdot \left[\mathrm{e}^{\frac{-V_S}{U_T}} - \mathrm{e}^{\frac{V_D}{U_T}} \right] \tag{2.5}$$

where I_S is the specific current given by

$$I_S = 2 \cdot n \cdot \beta \cdot U_T^2 \tag{2.6}$$

and $U_T = k \cdot T/q$ is the thermal voltage. In the saturation regime, i.e. when $V_D \gg V_S$, the expression reduces to

$$I_D = I_S \cdot \mathrm{e}^{\frac{V_G - V_T - n \cdot V_S}{n \cdot U_T}} \tag{2.7}$$

The transconductance in weak inversion saturation is given by

$$g_m = \frac{\partial I_D}{\partial V_G} = \frac{I_D}{n \cdot U_T} \tag{2.8}$$

2.1.3 *Moderate Inversion Regime*

In the moderate inversion regime, i.e. when the gate-to-source voltage is close to V_T, neither of the above expressions is accurate.

In the EKV model, the approach to model the transistor in moderate inversion involves selection of a smooth, continuous interpolation function F that consolidates the weak and strong inversion models into a single expression and provides a smooth transition between the two asymptotical regimes. The interpolation function must be chosen such that it becomes equal to (2.7) and (2.3) in deep weak and strong inversion, respectively.

The interpolation can be done on the drain current or on the transconductance, with the second method preferred over the first. In the latest version of the EKV model, version 3.0 described in [6], the interpolation function is defined as

$$g_m = \frac{I_D}{n \cdot U_T} \cdot \frac{1}{\frac{1}{2} + \sqrt{\frac{1}{4} + \frac{I_D}{I_S}}} \tag{2.9}$$

2.2 The MOS Differential Pair

The MOS differential pair is the primary building block of MCML logic gates. Here, contrary to most analog applications where it is used in its linear operating region, the differential pair is used as an all-or-nothing, or binary, current switch. Just as the combination of MOS transistors used as voltage switches in CMOS or pass-transistor logic styles enables the realization of logic functions, logic functions are realized in MCML by combining current switches into specific networks.

Analytical models of the MOS differential pair will now be derived in strong and weak inversion. These well-known models are accurate enough for small-signal modeling, and prove useful for getting an insight into the operation of the differential pair. However they do not allow to model the differential pair over a wide range of current as required for the modeling of MCML circuits. Therefore a transregional model will be presented next that accurately models the large-signal behavior of the differential pair.

2.2.1 *Strong Inversion Operation*

An MOS differential pair is depicted in Fig. 2.1. It is composed of two identical MOS transistors $M1$ and $M2$, with a common source connection S. A current source keeps a constant current I_{SS} flowing from the S node, therefore the sum of the drain currents of $M1$ and $M2$ is kept constant

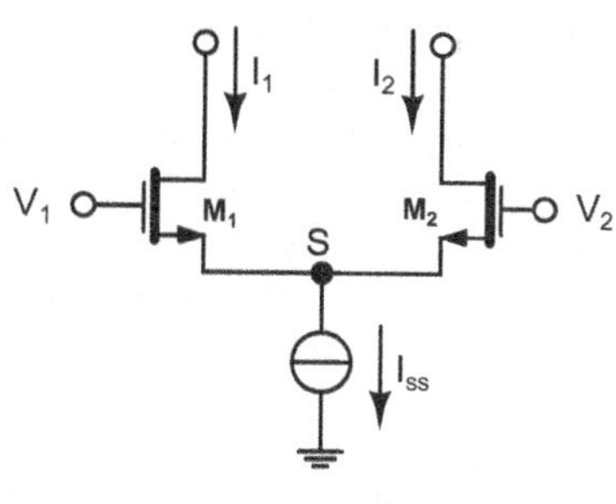

Fig. 2.1 A (N)MOS differential pair

$$I_1 + I_2 = I_{SS} \tag{2.10}$$

By summing the voltages around the loop $V_1 - V_2$, we obtain a second equation

$$V_1 - V_2 = V_{G1} - V_{G2} \tag{2.11}$$

Substituting the expression of the gate voltage as given by the strong inversion EKV model into Eq. (2.11) yields

$$V_1 - V_2 = (V_{T1} - V_{T2}) + \sqrt{\frac{2nI_1}{\beta_1}} - \sqrt{\frac{2nI_2}{\beta_2}} \tag{2.12}$$

Defining $V_{id} = (V_1 - V_2)$, the differential voltage at the input, and $\Delta I = (I_1 - I_2)$, the differential output current, using relation (2.10) then rearranging, we obtain

$$V_{id} = \sqrt{\frac{nI_{SS}}{\beta}}\left(\sqrt{1 + \frac{\Delta I}{I_{SS}}} - \sqrt{1 - \frac{\Delta I}{I_{SS}}}\right) \tag{2.13}$$

where it was assumed that $\beta_1 = \beta_2 = \beta$, and $V_{T1} = V_{T2}$, that is, both transistors are identical. This expression can be inverted to give the $\Delta I - V_{id}$ relationship

$$\frac{\Delta I}{I_{SS}} = \sqrt{\frac{\beta}{nI_{SS}}} V_{id} \sqrt{1 - \frac{1}{4}\frac{\beta}{nI_{SS}} V_{id}^2} \tag{2.14}$$

This relationship expresses the DC transfer curve for the circuit. It is important to notice that this equation relates the output *differential* current to the input *differential* voltage. The individual values of V_1 and V_2 do not appear—only their difference. Similarly, a differential current is produced as output; the average current, or common-mode current, is defined by the tail current I_{SS}.

Because the sum of $I_1$and I_2 must be equal to I_{SS}, it is clear that the maximum and minimum values of ΔI are limited to $\pm I_{SS}$. This happens when the tail current is entirely switched onto one of the two transistors, and the other is turned off. The value of V_{id} needed to accomplish this can be found by substituting $\Delta I = \pm I_{SS}$ in Eq. (2.13), yielding $V_{Ts} = \pm\sqrt{\frac{2nI_{SS}}{\beta}}$, where V_{Ts} denotes the switching threshold of the current switch.

The value of V_{Ts} reflects the ability of the paired transistors to drive the current. The larger β, the smaller V_{id} needs to be to switch an equal amount of current. This value is linked to the transconductance of the transistors, which reflects their current driving ability. Let us define the differential transconductance g_{md}, defined as the small-signal increase in output current caused by an increase in input voltage

$$g_{md} = \frac{\partial \Delta I}{\partial V_{id}} = \left(\frac{\partial V_{id}}{\partial \Delta I}\right)^{-1}$$

$$= \left[\sqrt{\frac{nI_{SS}}{\beta}} \cdot \frac{1}{I_{SS}} \cdot \left(\frac{1}{2\sqrt{1 + \frac{\Delta I}{I_{SS}}}} + \frac{1}{2\sqrt{1 - \frac{\Delta I}{I_{SS}}}} \right) \right]^{-1}$$

$$= \frac{2g_{md,0}}{\left(1 + \frac{\Delta I}{I_{SS}}\right)^{-\frac{1}{2}} + \left(1 - \frac{\Delta I}{I_{SS}}\right)^{-\frac{1}{2}}}$$

where

$$g_{md,0} = g_{md}|_{\Delta I=0} = \sqrt{\frac{\beta I_{SS}}{n}} = \frac{I_{SS}\sqrt{2}}{V_{Ts}} \tag{2.15}$$

is the differential transconductance at equilibrium ($V_{id} = 0$), and is equal to the transconductance of a single transistor biased at $I_D = I_{SS}/2$. Rewriting (2.14) in terms of $g_{md,0}$, we obtain

$$\frac{\Delta I}{I_{SS}} = \frac{g_{md,0}}{I_{SS}} V_{id} \sqrt{1 - \frac{1}{4}\left(\frac{g_{md,0}}{I_{SS}} V_{id}\right)^2} \tag{2.16}$$

The influence of transistor parameters is now explicit: their transconductance directly defines the slope of the $\Delta I - V_{id}$ curve and the whole transfer characteristic. It is therefore very practical, to normalize V_{id} to $\left(g_{md,0}/I_{SS}\right)^{-1}$ and ΔI to I_{SS}, as it is done in Fig. 2.2 which displays a plot of expression (2.16).

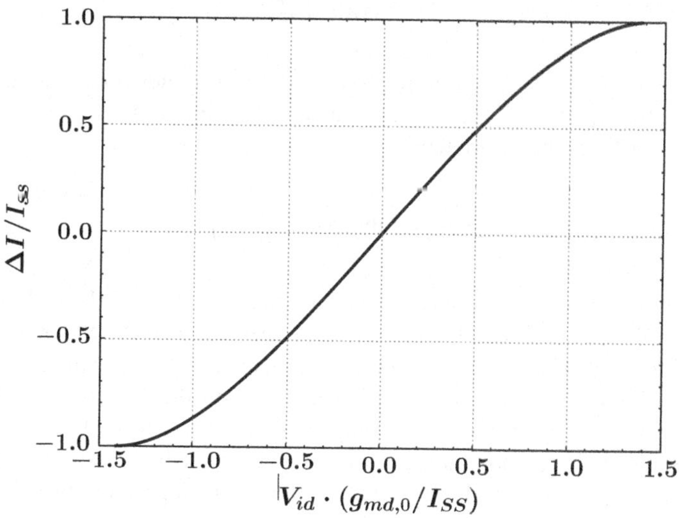

Fig. 2.2 DC transfer curve of the MOS differential pair in strong inversion

2.2.2 *Subthreshold Operation*

As we have seen previously, the gain in the differential pair is proportional to the transconductance of the transistors biased at $I_{SS}/2$. As we increase the transistor sizes in order to increase the gain, they will eventually enter subthreshold regime. When both transistors are in subthreshold regime, their drain currents grow exponentially with the gate-to-source voltage, according to the EKV model

$$I_D = I_S e^{\frac{V_G - V_T - n \cdot V_S}{n \cdot U_T}}$$

Following the same steps as for Eqs. (2.10)–(2.16), the $V_{id} - \Delta I$ relationship is found to be expressed as

$$V_{id} = nU_T \left[\ln \left(1 + \frac{\Delta I}{I_{SS}} \right) - \ln \left(1 + \frac{\Delta I}{I_{SS}} \right) \right]$$

The equation can be reversed to express $\frac{\Delta I}{I_{SS}}$, yielding

$$\frac{\Delta I}{I_{SS}} = \frac{e^{\frac{V_{id}}{nU_T}} - 1}{e^{\frac{V_{id}}{nU_T}} - 1} = \tanh \left(\frac{1}{2} \frac{V_{id}}{nU_T} \right) \tag{2.17}$$

In this regime, the transconductance has a different expression

$$g_{md} = \frac{g_{md,0}}{\cosh^2 \left(\frac{V_{id}}{2nU_T} \right)}$$

where $g_{md,0} = I_{SS}/\,(2 \cdot nU_T)$ is the differential transconductance at equilibrium, and is here also equal to the transconductance of a single transistor biased at $I_{SS}/2$. Therefore, (2.17) can be normalized as

$$\frac{\Delta I}{I_{SS}} = \tanh \left(V_{id} \frac{g_{md,0}}{I_{SS}} \right)$$

A few observations can be made regarding the operation of the differential pair in subthreshold region. First, the ratio of $g_{md,0}$ over I_{SS} is independent of transistor size. This means that the shape of the transfer curve (Fig. 2.3) cannot be changed by design. Moreover, $g_{md,0}$ is strongly dependent on the temperature. Therefore, in order to maintain a constant transconductance, the tail current should be varied proportionally to the temperature. Second, according to the model, the current will never be entirely switched: the differential pair is an imperfect current switch. In order to switch a given fraction of the current, the input differential voltage should be as large as $2nU_T \tanh^{-1} (\Delta I/I_{SS})$, which is about 5.3 times the thermal voltage (138 mV at room temperature) for a 99% switching.

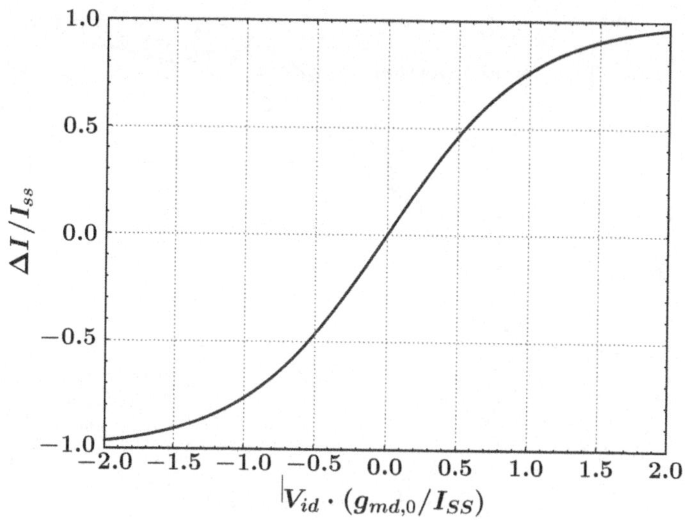

Fig. 2.3 Transfer curve of the differential pair in subthreshold regime

2.2.3 *Transregional Model*

So far, we have presented models of the differential pair both in strong and weak inversion. Both models give simple expressions to predict the shape of the transfer characteristic, which is a desirable property. However, they are both based on the strong assumption that b*oth transistors operate in the same regime*, and this assumption can only hold over a limited range.

In most of the cases, the large-signal operation of the differential pair cannot be analyzed accurately by considering a single mode of operation for the transistors—the exception being when it is operated in deep weak-inversion, over a range of currents where the subthreshold exponential $I_D - V_G$ relationship holds with good accuracy. When biased in strong (or moderate) inversion, the strong-inversion model will remain valid only as long as the gate-to-source voltages of both transistors remain large enough compared to V_T, or equivalently as long as the current remains high enough in both branches. This is the reason why a strong-inversion model can correctly predict the behavior of the differential pair in the central region of the transfer characteristic, which is close to linear, but fails to accurately model the regions where the curve is bending.

Therefore, in order to produce an accurate large-signal model valid over the whole operating range of the differential pair, both regimes should be considered. This is supported by observing the poor matching of strong- and weak-inversion models to the actual simulated transfer curves of differential pairs, as plotted in Fig. 2.4. As it can be concluded from the observation of these curves, the models are accurate for small-signal operation, when both transistors operate in the same

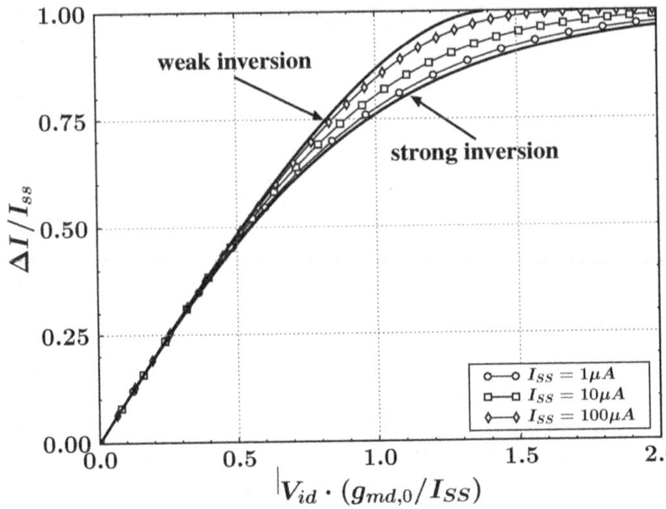

Fig. 2.4 Transfer curve of a differential pair in 90 nm CMOS at different current levels and the ideal strong- and weak-inversion models ($L = L_{\min}$, $W = 5 \times L$)

regime, however as V_{id} is increased, the real characteristics deviate from the expected ones and neither model offers acceptable accuracy. The actual value of ΔI lies somewhere between the strong inversion and the weak inversion expressions.

In order to model the large-signal behavior of the differential pair over a wide range of input voltages, we can adopt the interpolation approach of the EKV model. However, the interpolation function (2.9) is too complex to yield tractable expressions for hand analysis. Therefore, we will use the simpler expression

$$g_m(I_D) = \frac{g_s g_w}{g_s + g_w} \tag{2.18}$$

where g_s and g_w are the expressions of the transconductance in strong and weak inversion, respectively. This interpolation is valid under the assumptions that $g_w/g_s \to \infty$ when $I_D \to \infty$, and $g_w/g_s \to 0$ when $I_D \to 0$, which hold for MOSFETs. This is graphically represented in Fig. 2.5.

Then, since $g_m = dI_D/dV_G$, we can obtain V_G by integrating $1/g_m$ as follows:

$$V_G = \int \frac{\partial V_G}{\partial I_D} dI_D + C = \int \frac{1}{g_m} dI_D + C = \int \left(\frac{1}{g_s} + \frac{1}{g_w} \right) dI_D + C$$

$$= V_{G(strong)}(I_D) + V_{G(weak)}(I_D) + C \tag{2.19}$$

Now let us calculate the $V_{id} - \Delta I$ relationship for the differential pair using this expression of V_G

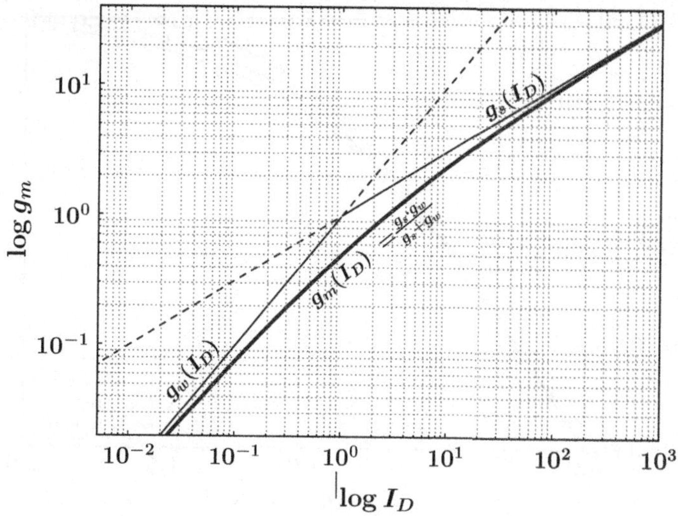

Fig. 2.5 Continuous interpolation of the transconductance from weak to strong inversion according to (2.18)

$$
\begin{aligned}
V_{id} &= V_G(I_1) - V_G(I_2) \\
&= \left[V_{G(strong)}(I_1) - V_{G(strong)}(I_2)\right] + \left[V_{G(weak)}(I_1) - V_{G(weak)}(I_2)\right] \\
&= V_{id(strong)}(\Delta I) + V_{id(weak)}(\Delta I)
\end{aligned}
\tag{2.20}
$$

where $V_{id(strong)}$ and $V_{id(weak)}$ are the $V_{id} - \Delta I$ transfer curves for the differential pair in strong and weak inversion, respectively. Inserting (2.16) and (2.17) into (2.20) and rearranging, we obtain

$$
\frac{V_{id}}{2nU_T} = \sqrt{\frac{I_{SS}}{2I_S}} \cdot \left(\sqrt{1 + \frac{\Delta I}{I_{SS}}} - \sqrt{1 - \frac{\Delta I}{I_{SS}}}\right) + \tanh^{-1}\left(\frac{\Delta I}{I_{SS}}\right) \tag{2.21}
$$

where the quantity $I_{SS}/(2I_S)$ in this expression is equal to the inversion factor at $I_D = I_{SS}/2$ as defined in the EKV model. The small-signal differential transconductance at equilibrium is given by

$$
g_{md,0} = \frac{I_{SS}}{2nU_T} \cdot \frac{1}{1 + \sqrt{\frac{I_{SS}}{2I_S}}} \tag{2.22}
$$

Note that the maximum transconductance is equal to the value in weak inversion, i.e. $I_{SS}/(2nU_T)$.

The good accuracy of this model can be assessed on the data plotted in Fig. 2.6, where the model is plotted against the same data as in Fig. 2.4 obtained

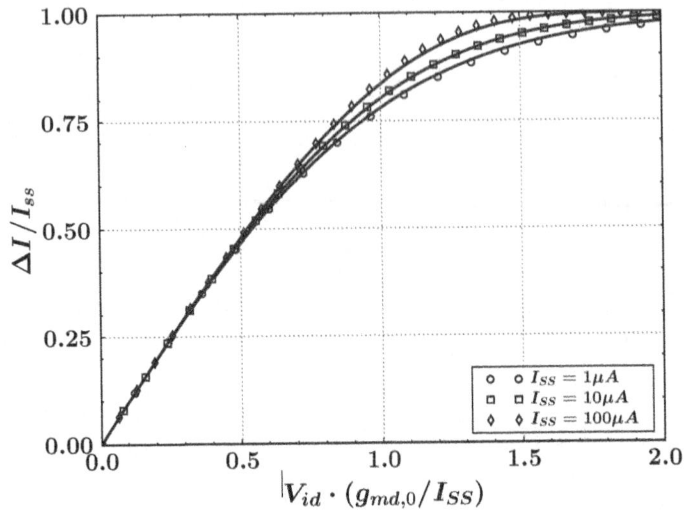

Fig. 2.6 Transfer curve of a differential pair in 90 nm CMOS at different current levels and the transregional model ($L = L_{\min}$, $W = 5 \times L$)

by simulation in a 90 nm CMOS process. For each current level, the differential transconductance $g_{md,0}$ has been extracted from the simulated data, by measuring the slope of the transfer curve in the linear region, and the resulting value used to calculate the I_S parameter in order to construct the analytical curves.

2.3 Single-Level MCML Logic Gate

The simplest MCML gate is built with a single level of differential pairs in the logic network. Only a single gate can be realized with one level, which is illustrated in Fig. 2.7. This gate realizes the function of a buffer or—since logically inverting a differential signal is done simply by swapping the two polarities—a logic inverter as well. It is composed of a single differential pair, producing a differential current from a differential input voltage. This differential current is converted into a differential output voltage, by means of load resistors, and the resulting differential voltage can be fed to the next logic stage. In this section, its operation will be studied in detail.

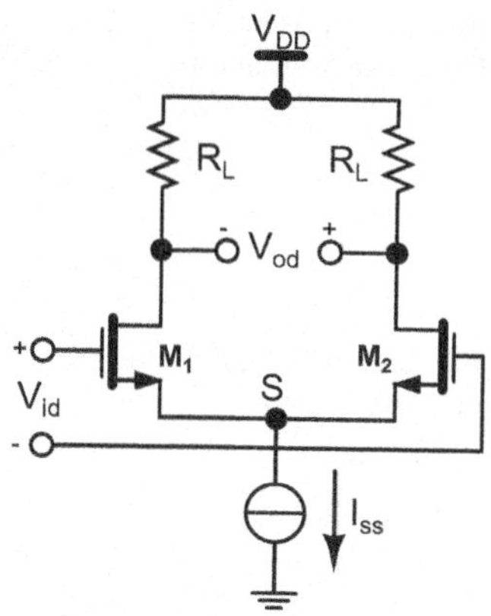

Fig. 2.7 MOS current-mode logic inverter/buffer

2.3.1 *Implementation of Load Devices*

Practically, the pull-up devices in an MCML gate can be implemented either as passive or active devices. In the first case, various flavors exist in modern VLSI technologies:

- diffused resistors are implemented by the parasitic resistance of low-doped silicon. Depending on the doping, they can offer sheet resistances as high as 1 kΩ/square, however they suffer from high parasitic capacitance due to the reverse biased pn-junction.
- polysilicon resistors are implemented with unsalicided strips of polysilicon. Typically, polysilicon resistors offer sheet resistances in the range 200–500 Ω/square, and up to 1 kΩ/square in processes offering high-resistive polysilicon. This type of resistance is more linear than diffused resistors and suffers less parasitic capacitance.

Passive resistors are inherently subject to process variations, due to doping, lithography, and etching. The tolerance on the absolute value of passive devices is typically as large as 20–30% of the nominal value.

Active resistors can be implemented with MOS devices operating in the linear region. They can offer an acceptable linearity if their V_{GS} is high and the voltage swing is small. Active resistors are naturally adjustable, by varying the bias voltage, so on-chip biasing can be implemented to compensate the process variations, and obtain very precise absolute values. Typical values for such passive resistors are in the range of 10–50 kΩ/square of channel dimensions (Fig. 2.8).

For very low current levels, active resistors with bulk biasing can be implemented to increase the resistivity, allowing up to several hundred of MΩ/square. This technique is explained in detail in [17, 18], and can be used for circuits biased in weak inversion, with very low current levels.

The choice of a type of device largely depends on the range of resistor values that are needed. For high values, passive components are growing up to unacceptable

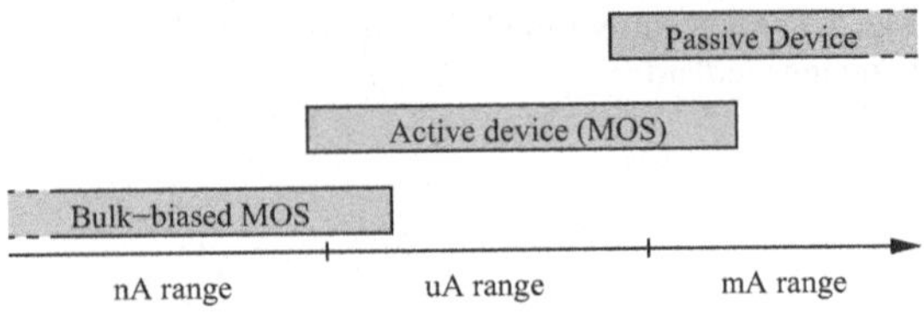

Fig. 2.8 Resistor devices and their range of usability

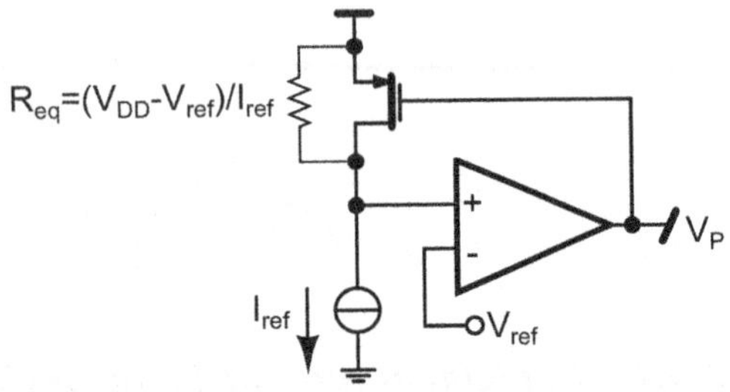

Fig. 2.9 On-chip adjustment of active resistor

sizes, while for low resistance values their dimensions and parasitics make them more efficient than active devices. Adjustment of the total resistance with passive devices can be achieved by circuit techniques, for example by combining active and passive devices in parallel, resulting in smaller overall dimensions than the use of an active device (Fig. 2.9).

In this work, we are focusing on MCML design with application to standard-cell type of circuits, which implies that the device sizes that we are considering are in the range of the technology minimum, and that the devices should be biased in strong or moderate inversion to allow high speed operation. For this type of application, active loads offer the best compromise, and in the remaining of this text we will assume that load devices are realized with PMOS transistors biased in the linear region, unless explicitly stated otherwise.

2.3.2 *DC Transfer Characteristic*

In order to obtain an accurate large-signal model for the $V_{od} - V_{id}$ relationship of the circuit in Fig. 2.7, we use the transregional model proposed in the previous sections for the transfer function of the differential pair.

Derivation The differential voltage V_{od} at the output of this circuit is given by

$$\begin{aligned} V_{od} &= (V_{DD} - R_L I_2) - (V_{DD} - R_L I_1) \\ &= R_L (I_1 - I_2) = R_L \Delta I \end{aligned} \tag{2.23}$$

Substituting ΔI as given by (2.21), we obtain the expression of the voltage transfer characteristic (VTC) as

$$\frac{V_{id}}{2nU_T} = \sqrt{\frac{I_{SS}}{2I_S}} \cdot \left(\sqrt{1+\frac{V_{od}}{V_{sw}}} - \sqrt{1-\frac{V_{od}}{V_{sw}}}\right) + \tanh^{-1}\left(\frac{V_{od}}{V_{sw}}\right) \tag{2.24}$$

where the *voltage swing* $V_{sw} = R_L I_{SS}$ is equal to the maximum voltage drop across the individual load resistors (when $\pm \Delta I = I_{SS}$). For this circuit, the *small-signal differential voltage gain* is given by

$$A_{vd} = g_{md,0} \cdot R_L = \frac{V_{sw}}{2nU_T} \cdot \frac{1}{1+\sqrt{\frac{I_{SS}}{2I_S}}} \tag{2.25}$$

using (2.22). This quantity can conveniently be substituted in the various expressions, with the desirable property of being directly measurable on the VTC as the slope at the origin. However, it encapsulates a dependency on V_{sw}, I_{SS}, and β which we regard as a fundamental design parameters. It is preferable when dealing with design problems to reason in terms of these parameters, that can be directly adjusted on the circuit and thus provide more useful insight into the tradeoffs. Therefore, we will exchangeably formulate the expressions in terms of *measurable* quantities when a comparison is required with experimental data, and in terms of *adjustable* parameters when discussing design decisions.

Approximation Expression (2.24) is a linear combination of two terms, corresponding, respectively, to the strong and weak inversion regimes. Because the weak inversion term accounts for most of the nonlinearity in the overall characteristic, the strong inversion term can be approximated by linearization with little loss of accuracy. This leads to the following simpler expression

$$\frac{V_{id}}{2nU_T} = \sqrt{\frac{I_{SS}}{2I_S}} \cdot \frac{V_{od}}{V_{sw}} + \tanh^{-1}\left(\frac{V_{od}}{V_{sw}}\right) \tag{2.26}$$

Validation Expressions (2.24) and (2.26) are plotted in Fig. 2.10 together with simulation results in a 90 nm CMOS technology. The simulation results are obtained with two different sizes for the transistors in the differential pair, at $I_{SS} = 10\,\mu\text{A}$, $V_{sw} = 400\,\text{mV}$, $T = 300\,\text{K}$ and with pure resistive loads. In addition, the curves resulting from simple weak and strong inversion models are included, highlighting the benefits of using a transregional model.

2.3.3 Noise Margin

The noise margin is an important parameter of any logic gate, characterizing its robustness to external perturbations, or equivalently its ability to provide a correct output in the presence of noise. The problem of quantifying the noise robustness

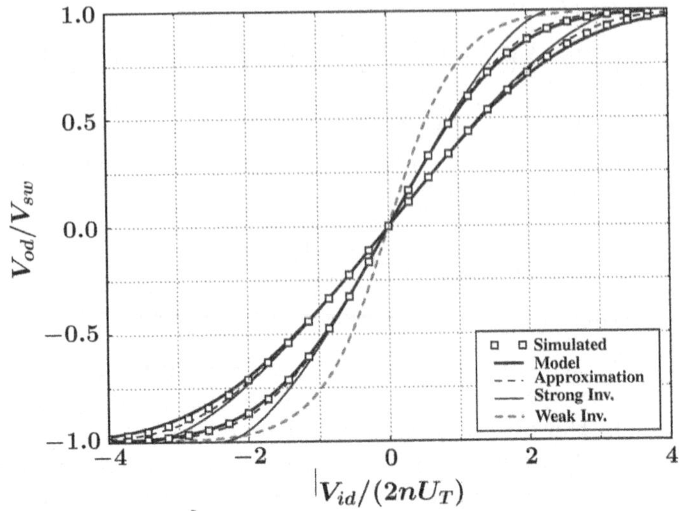

Fig. 2.10 Simulated transfer function of an MCML buffer in 90 nm CMOS technology, and the analytical model. Parameters are: $V_{DD} = 1.2\,\text{V}$, $I_{SS} = 10\,\mu\text{A}$, $V_{sw} = 400\,\text{mV}$, $T = 300\,\text{K}$

of a circuit is quite complicated, especially when considering a dynamic operation where the effect of noise is function of the duration and shape of the noise signals. For this reason, the most widely used criterion for quantifying noise robustness is the worst-case static noise margin. Worst-case static noise margins are evaluated by considering a quasi-static situation, where the duration of the noise is very long compared to the response time of the gates, and a worst case scenario where an equal amount of noise is applied at the input of each logic gate along an infinitely long chain. The noise margins are then defined as the maximum noise amplitude that does not perturb the logic state of the circuit.

As discussed in [10], the noise margin can be represented graphically by drawing the voltage transfer curve (VTC) of a gate, mirroring it along the $y = x$ line and drawing the maximum-sized square that fits in the area between the two curves (Fig. 2.11). Furthermore, as pointed out in [11], the V_{IH}, V_{OH}, V_{IL}, and V_{OL} points as defined in the figure are the coordinates of the points where the derivative of the VTC is equal to one.

Derivation Differentiating (2.26) implicitly with respect to V_{od}, we can express the derivative of the VTC for an MCML buffer as

$$\frac{\partial V_{id}}{\partial V_{od}} = \frac{2nU_T}{V_{sw}} \cdot \left[\sqrt{\frac{I_{SS}}{2I_S}} + \frac{1}{1 - \left(\frac{V_{od}}{V_{sw}}\right)^2} \right] \tag{2.27}$$

Equating to unity and solving, we obtain V_{OH}

Fig. 2.11 Graphical representation of static noise margins

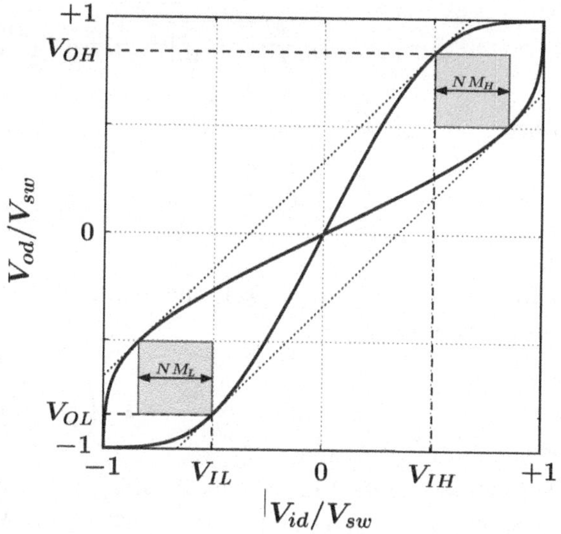

$$V_{OH} = V_{sw} \cdot \sqrt{1 - \frac{1}{\frac{V_{sw}}{2nU_T} - \sqrt{\frac{I_{SS}}{2I_S}}}} \tag{2.28}$$

Replacing back into (2.26), we obtain V_{IH}.

$$V_{IH} = 2nU_T \cdot \left[\sqrt{\frac{I_{SS}}{2I_S}} \cdot \sqrt{1 - \frac{1}{\frac{V_{sw}}{2nU_T} - \sqrt{\frac{I_{SS}}{2I_S}}}} + \tanh^{-1}\left(\sqrt{1 - \frac{1}{\frac{2nU_T}{V_{sw}} - \sqrt{\frac{I_{SS}}{2I_S}}}} \right) \right] \tag{2.29}$$

Next, NM_H is given by $|V_{OH} - V_{IH}|$. By symmetry, we find that $NM_H = NM_L = NM$, therefore

$$NM = \left(V_{sw} - 2nU_T \cdot \sqrt{\frac{I_{SS}}{2I_S}} \right) \sqrt{1 - \frac{1}{\frac{V_{sw}}{2nU_T} - \sqrt{\frac{I_{SS}}{2I_S}}}} - 2nU_T \cdot \tanh^{-1}\left(\sqrt{1 - \frac{1}{\frac{V_{sw}}{2nU_T} - \sqrt{\frac{I_{SS}}{2I_S}}}} \right) \tag{2.30}$$

Approximation This expression is not convenient as it cannot be reversed in order to express design parameters as a function of the noise margin. Let us rewrite it as

$$\frac{NM}{2nU_T} = \xi\sqrt{1-\frac{1}{\xi}} - \tanh^{-1}\sqrt{1-\frac{1}{\xi}} \tag{2.31}$$

where

$$\xi = \frac{V_{sw}}{2nU_T} - \sqrt{\frac{I_{SS}}{2I_S}} = 1 + \frac{V_{sw}}{2nU_T}\left(1-\frac{1}{A_{vd}}\right) \tag{2.32}$$

Expression (2.31) can be well approximated over the range of interest by the following formula

$$\frac{NM}{2nU_T} \approx \xi\cdot\left(1-\frac{1}{\xi}\right)^2 \tag{2.33}$$

Validation Figure 2.12 plots the resulting expression against the values obtained by simulation using a 90 nm CMOS technology. As it can be seen, the analytical formula agrees well with the simulated data. It tends to slightly overestimate the noise margin when the pair is strongly inverted, i.e. for low values of A_{vd} and large values of V_{sw}, because of the linear approximation of the strong inversion characteristic. For larger values of V_{sw}, the transistors may enter linear region when their drain voltage is low—that is, when V_{id} is large—effectively reducing

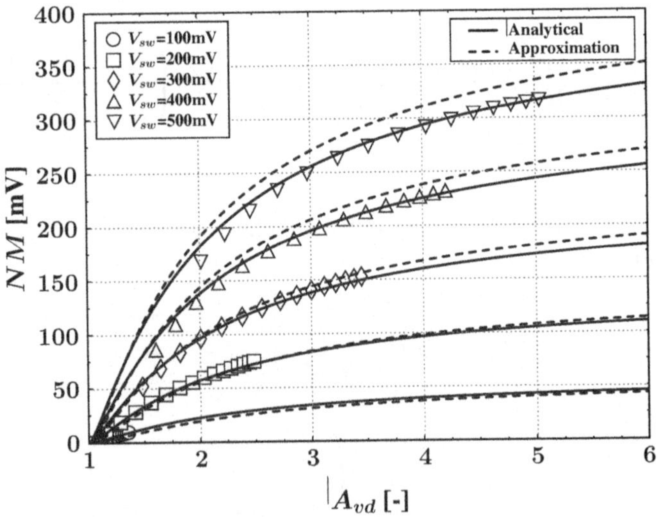

Fig. 2.12 Simulated noise margin of an MCML buffer in 90 nm CMOS for different values of the voltage swing, together with the analytical expression (2.30) and the approximation (2.33). Simulation parameters are: $V_{DD} = 1.2\,\text{V}$, $I_{SS} = 10\,\mu\text{A}$, $T = 300\,\text{K}$

their transconductance and thus flattening the transfer curve. This has the effect of reducing the noise margin, and is also not taken into account in the analysis.

Expression (2.33), also displayed in Fig. 2.12, gives a very good approximation over a broad range of operating points. Several more simple expressions can be found, by considering only the weak or strong inversion behavior, however these expressions are poorly accurate, or accurate only within a limited range of operating points.

The weak-inversion limit is found by setting $I_S \gg I_{SS}$ in the previous expression, yielding

$$\frac{NM_w}{V_{sw}} = \sqrt{1 - \frac{1}{A_{vd}}} - \frac{1}{A_{vd}} \cdot \tanh^{-1}\sqrt{1 - \frac{1}{A_{vd}}} \quad (2.34)$$

Conversely, the noise margin considering only the strong-inversion behavior is found by setting $I_S \ll I_{SS}$, yielding

$$\frac{NM_{s,1}}{V_{sw}} = 1 - \frac{1}{A_{vd}} \quad (2.35)$$

corresponding to a velocity saturation index [14, 15] of $\alpha = 1$, because this assumption was made to derive the initial formula (by linearizing the strong inversion characteristic). The expression for the case where $\alpha = 2$ can be found by following the same approach as previously, and was first calculated in [5] as

$$\frac{NM_{s,2}}{V_{sw}} = \frac{\sqrt{4A_{vd}^2 - 1 - \sqrt{8A_{vd}^2 + 1}}}{A_{vd}^2\sqrt{2}} \left(\frac{\sqrt{4A_{vd}^2 + 1 + \sqrt{8A_{vd}^2 + 1}}}{2\sqrt{2}} - 1 \right) \quad (2.36)$$

As suggested in [2], this expression can be approximated by assuming $V_{OH} = V_{sw}$ and $A_{vd} \gg 1/\sqrt{8}$, yielding a simpler formula

$$\frac{NM_{s,3}}{V_{sw}} = 1 - \frac{\sqrt{2}}{A_{vd}}\sqrt{1 - \frac{1}{A_{vd}\sqrt{2}}} \quad (2.37)$$

or, by approximating even further

$$\frac{NM_{s,4}}{V_{sw}} = 1 - \frac{\sqrt{2}}{A_{vd}} \quad (2.38)$$

The latter expression is found to be of the same form as (2.35), and by extension the authors in [1] proposed to express the noise margin for arbitrary values of α as

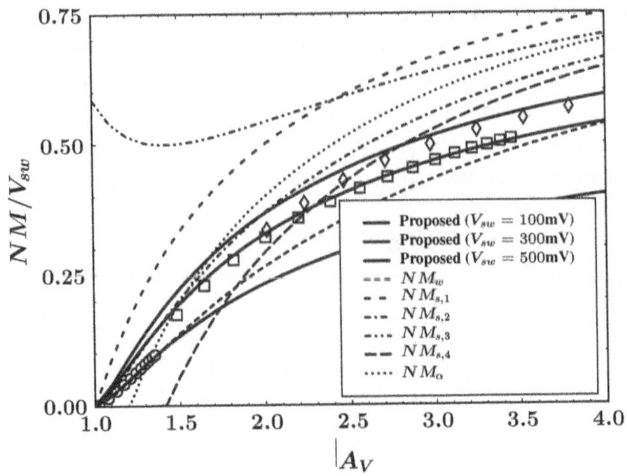

Fig. 2.13 Comparison of different noise margin formulas. The symbols denote simulation results from Fig. 2.12, at three different V_{sw}. The bold lines show the theoretical noise margin values from expression (2.30). The different thin lines show the resulting values from expressions (2.34)–(2.39)

$$\frac{NM_\alpha}{V_{sw}} = 1 - \frac{\gamma}{A_{vd}} \tag{2.39}$$

where γ is a constant between 1 and $\sqrt{2}$, reflecting the effect of velocity saturation.

In order to compare all these different expressions, verify their accuracy and assess their range of validity, they are all displayed on the same graph in Fig. 2.13 together with the simulated data from Fig. 2.12.

By inspecting the matching of the different curves to the simulated data in Fig. 2.13, it appears that all approximations (2.34)–(2.39) suffer from an important degradation in accuracy, compared to expression (2.30). The weak-inversion approximation (2.34) is accurate at lower values of V_{sw}. Using the long-channel strong-inversion expression (2.36) is reasonable for small values of A_{vd}, however this expression is not simple enough to be used in calculations. For larger values ($A_{vd} > 2$) the approximations (2.38) and (2.39) are actually closer to the simulated data, but they have the undesirable drawback of altering the x-axis intercept at $A_{vd} = 1$. Those three expressions are nevertheless preferable over (2.35), (2.37) which are rather inaccurate.

2.3.4 *Logic Levels*

When considering a chain of (identical) logic gates, the voltage levels will tend to stabilize along the chain to the values which satisfy the relation $V_{in} = V_{out}$. There should be two such values for binary logic, one for each logic level, and these values are termed the *stable logic levels*, denoted V_H and V_L.

Derivation In the case of an MCML buffer with a transfer characteristic given by (2.24), symmetry implies that $V_H = -V_L = V_{HL}$, and the logic levels must satisfy

$$\frac{V_{HL}}{2nU_T} = \sqrt{\frac{I_{SS}}{2I_S}} \cdot \left(\sqrt{1 + \frac{V_{HL}}{V_{sw}}} - \sqrt{1 - \frac{V_{HL}}{V_{sw}}}\right) + \tanh^{-1}\left(\frac{V_{HL}}{V_{sw}}\right) \tag{2.40}$$

Unfortunately, this transcendental equation cannot be solved analytically. However, it can be reversed to obtain parameter values that result in a given logic level specification. Defining $V_{HL}/V_{sw} = \alpha$, we can write

$$\xi = \frac{V_{sw}}{2nU_T}\left(1 - \frac{\alpha}{\sqrt{1+\alpha} - \sqrt{1-\alpha}}\right) + \frac{\tanh^{-1}(\alpha)}{\sqrt{1+\alpha} - \sqrt{1-\alpha}} \tag{2.41}$$

where ξ has the definition given in (2.32)

$$\xi = \frac{V_{sw}}{2nU_T} - \sqrt{\frac{I_{SS}}{2I_S}} \tag{2.42}$$

Validation The relationship between the stable logic levels and the ξ value is plotted in Fig. 2.14 from simulations results in 90 nm CMOS technology at different voltage swings, ranging from 100 to 500 mV, and various device sizes for the transistors in the differential pair. On the same graph, the theoretical value as given by (2.42) is plotted with dotted lines.

Some important conclusions can be drawn from the inspection of this graph. Firstly, note that a given value of ξ corresponds to a unique value of NM, as given by (2.31). Therefore, at a given noise margin, gates with a lower voltage swing exhibit better logic levels. This is easily understood when considering that, for lower voltage swings, a differential pair with larger transconductance (resulting in a larger voltage gain) need to be used in order to compensate for the loss in noise margin. Secondly, the graph shows that good logic levels cannot be obtained at very low voltage swings. In fact, for a given value of V_{sw}, there is a maximum value of ξ given by

$$\xi_{\max} = \frac{V_{sw}}{2nU_T}$$

obtained when $I_S \gg I_{SS}$, i.e. when the transistor sizes in the differential pair grow to infinity. Therefore, for a specified value of α, one can deduce a lower bound on the voltage swing through (2.42), resulting in

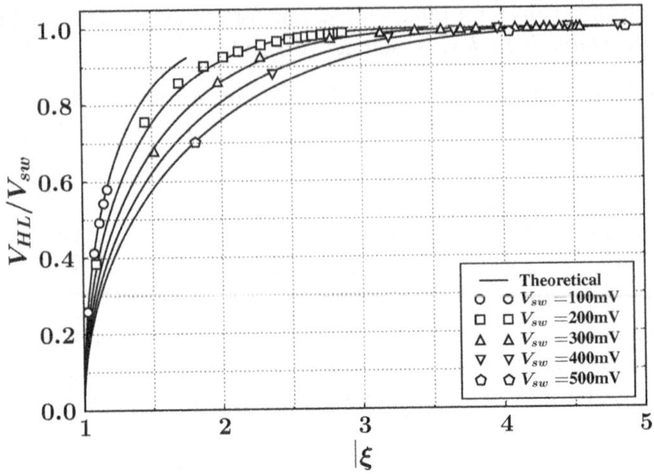

Fig. 2.14 Stable logic levels in an MCML gate. Symbols denote simulation results in a 90 nm CMOS process, for voltage swings ranging from 100 to 500 mV; bold lines represent the theoretical value given in Eq. (2.42)

$$V_{sw,\min} = 2nU_T \cdot \frac{\tanh^{-1}(\alpha)}{\alpha} \tag{2.43}$$

According to this result, the minimum allowable voltage swing to obtain a relative logic level of 99% is ~170 mV at room temperature, and rises to ~220 mV at $T =$ 400 K. In practice, unless I_{SS} is extremely low, the transistors need to be biased in moderate inversion in order to keep reasonable transistor sizes. This will further increase the minimum voltage swing to higher values.

2.3.5 *Dynamic Operation*

In order to achieve efficient design of MCML gates, it is important to understand their dynamic operation, that is, how they behave during switching events and how this behavior depends on their parameters. During the switching in an MCML gate, the two complementary outputs undergo an opposite change in voltage—one output being pulled low through the tail current of the differential pair while the second is pulled up to the supply voltage through the load resistor. Due to time constants in the system, these processes are not instantaneous and cause transition delay from input to output. But delay is not the only reason of analyzing the dynamic operation of a gate, for during the switching, multiple transient current paths can exist between supply and ground which may sum up as current spikes in the supply.

Figure 2.15 depicts an MCML buffer gate with the relevant parasitic capacitances at the input and output nodes. Assuming that the circuit is symmetrical by design and

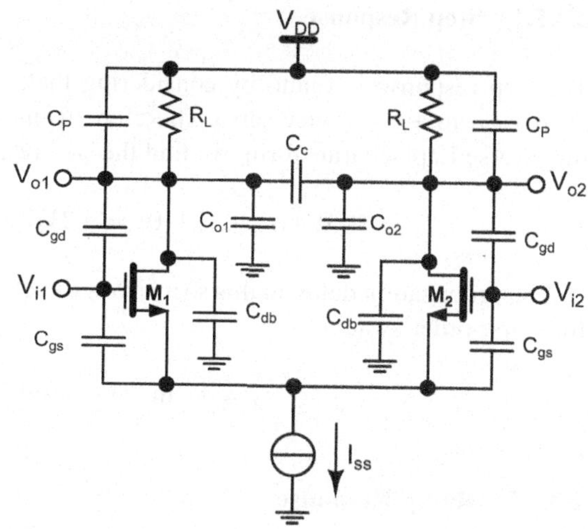

Fig. 2.15 Parasitic capacitances in an MCML buffer

by layout, the capacitance contributions are equal on both sides. The time constant at the output nodes is determined by the parasitics of the transistors (C_{db}, the drain to bulk diffusion capacitance, and C_{gd}, the gate-to-drain capacitance), the parasitics due to the pull-up load device denoted here by C_P, and the external parasitics C_o and C_c due to wiring and input capacitance of the driven gates. All together, and taking into account the Miller effect, the total load capacitance can be estimated as

$$C_{tot} = 2C_{gd} + C_{db} + C_P + C_o + 2C_C$$

By writing Kirschoff's current law at both output nodes, we obtain the following differential equations

$$\tau \cdot \frac{dV_{o1,2}(t)}{dt} + V_{o1,2}(t) - \left[V_{DD} - R_L \frac{I_{SS} \pm \Delta I(t)}{2} \right] = 0 \tag{2.44}$$

Subtracting one from the other, the differential equation governing the output differential voltage is found as

$$\tau \cdot \frac{dV_{od}(t)}{dt} + V_{od}(t) - R_L \Delta I(t) = 0 \tag{2.45}$$

where τ is the time constant of the first-order system and is equal to $\tau = R_L C_{tot}$. After taking the Laplace transform and inverting, we get

$$V_{od}(s) = \frac{R_L \cdot \Delta I(s)}{1 + \tau \cdot s} \tag{2.46}$$

2.3.5.1 Step Response

The step response is found by considering that, during full switching, ΔI varies from $-I_{SS}$ to $+I_{SS}$ or vice versa. Therefore, using $\Delta I(s) = (\pm 2I_{SS})/s$, and taking the inverse Laplace transform, we find the step response as

$$V_{od}(t) - V_{od}(0) = \pm 2V_{sw}\left(1 - e^{-t/\tau}\right) \tag{2.47}$$

The propagation delay in this scenario is therefore equal to the well-known value for a first-order system

$$t_d = \tau \cdot \ln(2) \approx 0.69 R_L C_{tot} \tag{2.48}$$

2.3.5.2 Ramp Response

A step input is only valid in limit case where the input signal is changing very fast compared to the output signal. In real situations, the input and input slopes are usually of the same order of magnitude, and the propagation delay depends on the shape of the input signal. To model this effect, we can approximate the input waveform as a linear ramp. Because the transfer characteristic of the differential pair is fairly linear over a broad range of input voltage, we can approximate it with a linear relationship as

$$\frac{\Delta I}{I_{SS}} = \begin{cases} g_{md,0}V_{id} & \text{if } \left|\frac{V_{id}}{V_{sw}}\right| < \frac{1}{A_{vd}} \\ \pm 1 & \text{otherwise} \end{cases} \tag{2.49}$$

where the current saturates at $\pm I_{SS}$ when the input is large enough to switch the entire tail current. Instead of modeling the saturation of the transfer characteristic, it is simpler to clip the input waveform. Therefore, we will consider an input wave described as

$$V_{id}(t) - V_{id}(0) = 2\frac{V_{sw}}{A_{vd}} \cdot \frac{1}{\tau_i'} \cdot \left(t \cdot \varepsilon(t) - \left(t - \tau_i'\right) \cdot \varepsilon\left(t - \tau_i'\right)\right) \tag{2.50}$$

where τ_i' is the time needed for the input to rise from $-V_{sw}/A_{vd}$ to $+V_{sw}/A_{vd}$. Note that the time needed for the real input to rise from $-V_{sw}$ to $+V_{sw}$ is larger by a factor A_{vd}, therefore $\tau_i' = \tau_i/A_{vd}$ (Fig. 2.16). The output waveform is then given by

$$\begin{aligned} V_{od}(t) - V_{od}(0) = {} & \varepsilon(t)\left\{2\frac{V_{sw}}{\tau_i'}\left[t + \tau\left(e^{-t/\tau} - 1\right)\right]\right\} \\ & -\varepsilon\left(t - \tau_i'\right)\left\{2\frac{V_{sw}}{\tau_i'}\left[\left(t - \tau_i'\right) + \tau\left(e^{-(t-\tau_i')/\tau} - 1\right)\right]\right\} \end{aligned} \tag{2.51}$$

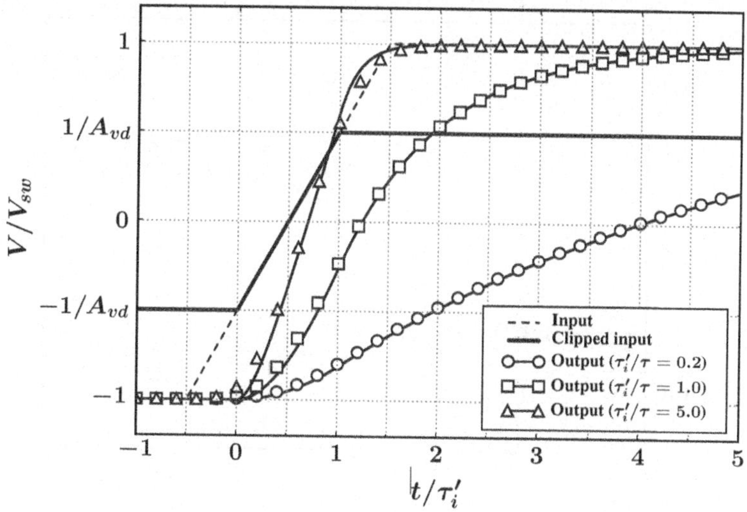

Fig. 2.16 Ramp response of an MCML buffer. Symbols denote simulation results with a 90 nm CMOS technology ($I_{SS} = 30\,\mu A$, $A_V = 2$)

By calculating the time point where $V_{od}(t) - V_{od}(0) = V_{sw}$, and subtracting $\tau_i'/2$, the expression for the delay is found. The resulting formula depends whether the output reaches the mid-point before or after the input saturates, which is function of the ratio τ_i/τ

$$\frac{t_d}{\tau} = \begin{cases} \frac{1}{2}\cdot\frac{\tau_i'}{\tau} - \ln\left(\frac{1}{2}\cdot\frac{\tau_i'/\tau}{1-e^{-\tau_i'/\tau}}\right) & \text{if } \frac{\tau_i'}{\tau} < 1.59362 \\ 1 + W\left[-e^{-\left(1+\frac{1}{2}\cdot\frac{\tau_i'}{\tau}\right)}\right] & \text{otherwise} \end{cases}$$

where W is known as *Lambert's W function* [3, 4, 7], and the constant 1.59362 is the approximate numeric value of $2 + W\left(2e^{-2}\right)$. Note that, for $\tau_i' \ll \tau$, the delay tends towards $\tau \ln(2)$, which is the value obtained with a step input, while for $\tau_i' \gg \tau$ it tends towards τ. Thus, the delay increase due to a very slowly changing input is at the most of about 30%, which is rather low (Fig. 2.17).

2.3.5.3 Power Supply Noise

During the event of switching, the current drawn from the power supply can vary due to the transient current paths charging and discharging the load capacitances. By summing the currents in both load devices, we get the total supply current as

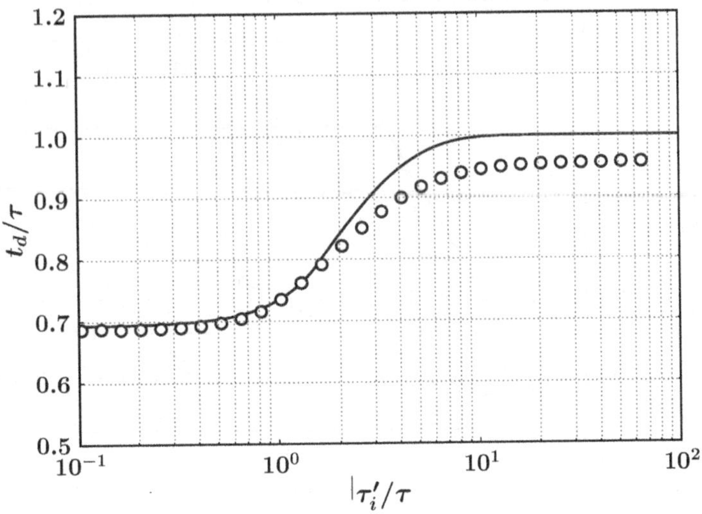

Fig. 2.17 Propagation delay of an MCML buffer as a function of the input rise time. Symbols denote simulation results with a 90 nm CMOS technology

$$I_{supply} = I_{SS} + C_L \frac{\mathrm{d}}{\mathrm{d}t}\left[V_{o1}(t) + V_{o2}(t)\right] \tag{2.52}$$

Therefore, power supply noise can result if the common-mode output voltage $V_{cm} = (V_{o1} + V_{o2})/2$ varies during the switching. By summing the differential equations for V_{o1} and V_{o2} as given by (2.44), then dividing by 2, we obtain V_{cm} as

$$\tau \cdot \frac{\mathrm{d}V_{cm}(t)}{\mathrm{d}t} + V_{cm}(t) = V_{DD} - \frac{V_{sw}}{2}$$

where both $R_L \cdot \Delta I$ component have cancelled out. Obviously, the common mode is thus constant and equal to the supply voltage minus half of the voltage swing. Therefore, the generated supply noise is null; it is important to realize, though, that this is possible only because the transient components on both output nodes cancel each other perfectly, and that any asymmetry will invalidate this condition and result in switching noise.

2.4 Multi-Level MCML Logic Gates

Complex logic functions are realized in MCML by stacking differential pair stages to build logic networks. Depending on the state of inputs applied to such a network, the tail current of the gate will be steered to one output node or the other, producing

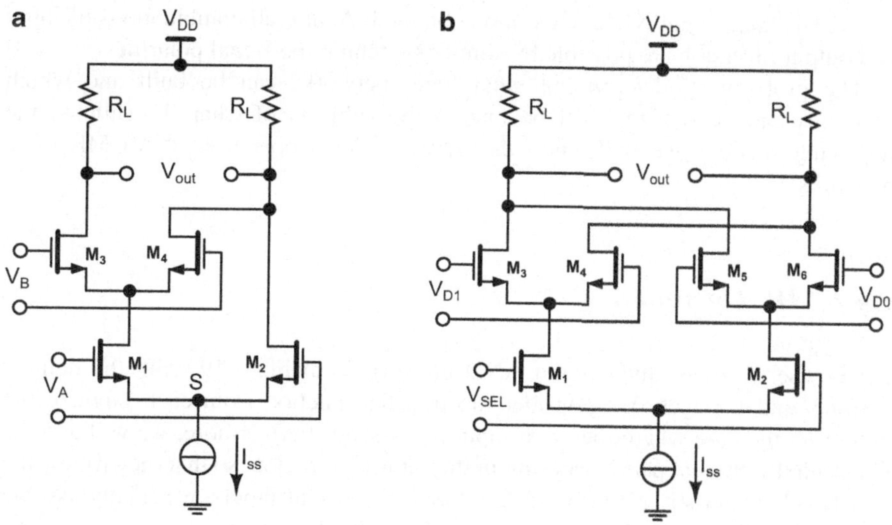

Fig. 2.18 (**a**) 2-Input AND/OR. (**b**) 2-to-1 multiplexer

a voltage drop at this node and effectively establishing the polarity of the differential output. Examples of logic networks are given in Fig. 2.18a, b.

The logic network implemented in Fig. 2.18a works as follows: when both inputs are in such a polarity that the differential pairs are steering current to their left branch, the current flows to the left output node, effectively pulling it low while the right output node is pulled high. In all other cases, the current is steered to the right output node. Polarities for inputs and output are intentionally not written on the picture, since all can be inverted at no cost, effectively realizing a logic inversion of an input or output: the same logic network will then be able to realize all possible variants obtained by inverting one or more of the inputs and/or output, without any change. Deciding on a polarity for inputs such that M_1 and M_3 are conducting when V_A and V_B are at the logic '1' state, and for the output such that it is at logic '1' state when M_5 is pulling low, we can write the following truth-table

V_A	V_B	V_{out}
0	0	0
0	1	0
1	0	0
1	1	1

and therefore, the gate with the chosen polarities implements the 2-input AND function. By switching polarities of both inputs and the output, the 2-input OR function can be realized. All other combinations, including NAND and NOR, are also possible with the same gate.

Similar analysis for the gate depicted in Fig. 2.18b reveals that the resulting logic function is $Y = S \cdot D_1 + \overline{S} \cdot D_0$, corresponding to the 2-to-1 multiplexer. By assigning

$D_0 = \overline{D_1}$, the 2-input XOR function is obtained. Again, all combinations of input and output inversion are possible by simply switching the signal polarities.

The problem of determining what logic networks can be built and which functions can be realized will be one of the subjects of Chap. 4, while in the following sections we will study the transistor-level operation of MCML logic networks.

2.4.1 DC Operation

Let us use as a case the 2-input AND network from Fig. 2.19, with the defined currents and voltages. We consider the transfer function from each input to the output, in the case where the other input is constant. Furthermore, we will assume that each differential pair is working in the saturation region, so that they follow the transfer characteristic given by (2.24). For the constant inputs, we finally assume that the pairs behave as ideal current switches, switching only a fraction γ of their tail current, where γ is related to the stable logic levels as defined in Sect. 2.3.4 through

$$\gamma = \frac{1+\alpha}{2} \tag{2.53}$$

Considering input A, we assume that V_B is positive, so that $I_3 = \alpha \cdot I_1$ and $I_2 = (1-\alpha) \cdot I_1$. The output voltage is given by

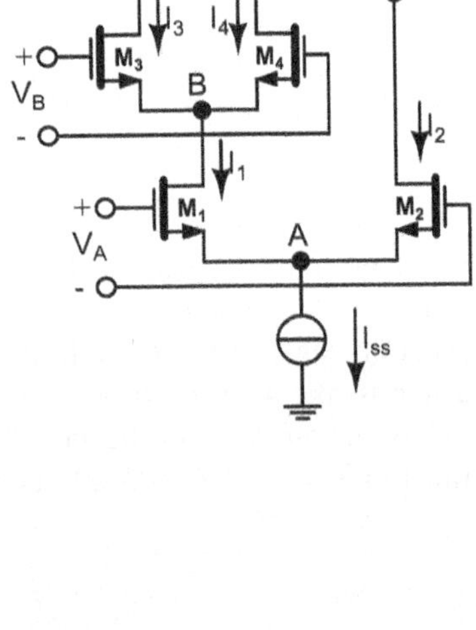

Fig. 2.19 DC voltages and currents in the AND2 logic network

$$\begin{aligned} V_{od} &= R_L \cdot (I_5 - I_6) \\ &= R_L \cdot [I_3 - (I_4 + I_2)] \\ &= R_L \cdot [\gamma \cdot I_1 - ((1-\gamma) \cdot I_1 + I_2)] \\ &= R_L \cdot [2\gamma \cdot I_1 - I_{SS}] \end{aligned}$$

Normalizing by dividing both sides by $R_L \cdot I_{SS}$, we obtain

$$\begin{aligned} \frac{V_{od}}{V_{sw}} &= (\gamma - 1) + \gamma \frac{\Delta I_{12}}{I_{SS}} \\ &= \frac{1+\alpha}{2} \cdot \frac{\Delta I_{12}}{I_{SS}} + \left(\frac{1+\alpha}{2} - 1\right) \end{aligned} \tag{2.54}$$

where

$$\Delta I_{12} = I_1 - I_2 \tag{2.55}$$

is the transfer function of the single differential pair as given by (2.21). We can make the following observations:

1. The transfer function is offset vertically by a factor $(1+\alpha)/2 - 1$
2. The transfer function is scaled by a factor $(1+\alpha)/2$
3. The logic levels are asymmetric

These effects are due to the imperfect current switching in the M_3–M_4 pair, and imply that the logic levels—and thus, the noise margin—are degraded in the multi-level gates when compared to a single-level gate with identical transistor dimensions.

When the transfer function is sufficiently flat at the crossing points, that is, for large enough values of α, the logic levels approximately scale according to (2.54). Substituting $\Delta I_{12} = \alpha$ results in

$$\alpha_2 \approx (1+\alpha) \cdot \frac{1+\alpha}{2} - 1 \tag{2.56}$$

Now, considering input B, similar calculations show that the transfer characteristic is affected in the same way. More generally, considering a gate with N levels of stacked differential pairs, in the worst case, the transfer function will be scaled by a factor of γ^{N-1} and offset by a factor $\gamma^{N-1} - 1$, resulting in a worst-case logic level of

$$\alpha_N \approx (1+\alpha) \cdot \left(\frac{1+\alpha}{2}\right)^{N-1} - 1 \tag{2.57}$$

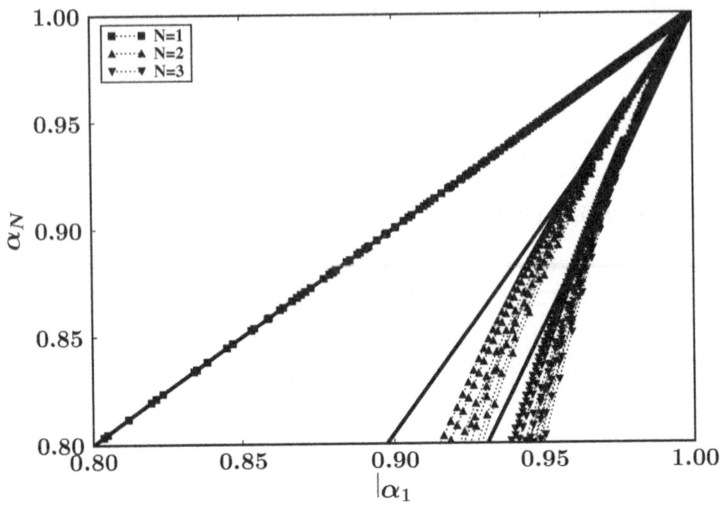

Fig. 2.20 Logic level degradation in multi-level MCML gates. The plain lines represent formula (2.57) for $N = 1, 2, 3$. The dotted lines denote simulation results obtained in a 90 nm CMOS technology, for AND gates with 1–3 inputs, with $I_{SS} = 100\,\mu A$, at various voltage swings and transistor sizes

Though this analysis is very simplistic, it allows to understand the effects involved. This is validated by the simulation results presented in Fig. 2.20. In this figure, expression (2.57) is plotted in plain lines for a gate with $N = 1, 2, 3$ levels of stacked differential pairs. The dotted lines represent simulation results, where the logic level corresponding to the N different inputs of each gate and four different values of V_{sw} are all superimposed. The results show that, for large values of α, the resulting logic level tends toward the curve given by (2.57).

The conclusion that can be drawn from these considerations is that, in order to design robust multi-level gates, the differential pairs should be able to efficiently switch the tail current. Moreover, with a higher number of levels, the current switching is less efficient and therefore the differential pairs should be sized larger. In practice, this implies that each additional level of stacked logic incurs additional costs in terms of area and speed.

As a final remark, it should be noted that it was assumed that all differential pairs are operating in saturation. In practice, this is possible only if the inputs to pairs at the different levels are level-shifted accordingly; this is discussed in more detail in the next section.

2.4.2 Common-Mode Input Level and Level Shifting

In a multi-level MCML network, differential pairs that are located on the highest level have their drain connected to the output nodes and the load devices, and their source, when no switching is occurring on lower level inputs, is driven by a constant current. Therefore, they operate exactly as if they were single-level, and their characteristics are the same as that of a buffer as analyzed in the previous section.

The same is not true, however, for a pair located on a lower level. In the network of Fig. 2.19, the M_1 transistor has its drain voltage equal to the source voltage of the M_3–M_4 differential pair. In a static condition, assuming that one of M_3 or M_4 has a gate voltage of V_{DD} and is conducting all the current, the source voltage V_B can be written as $V_B = (V_{DD} - V_T)/n - V_{ov}(I_1)$, where V_{ov} is the overdrive voltage required to drive I_1. Therefore, the condition for M_1 being saturated is written as

$$\frac{V_{DD} - V_T}{n} - V_{ov}(I_1) - \frac{V_1 - V_T}{n} \geq 0$$

or

$$V_1 \leq V_{DD} - V_{ov}(I_1) \tag{2.58}$$

Obviously, for this condition to remain true at all times, the input common mode-level of the M_1–M_2 input pair must be lowered by an amount equal to $V_{ov}(I_{SS})$. This can be achieved by inserting source–followers at the inputs as depicted in Fig. 2.21a. In this configuration, both inputs V_1 and V_2 are shifted down by an amount equal to $V_T + V_{ov}$, depending on the size of the transistors in the level shifters and their tail current. Another option is to insert source followers at the output of the gates, as displayed in Fig. 2.21b. The advantage of this configuration is that a single source–follower can provide a level-shifted signal to multiple gate inputs.

However, using level shifting has a number of important drawbacks:

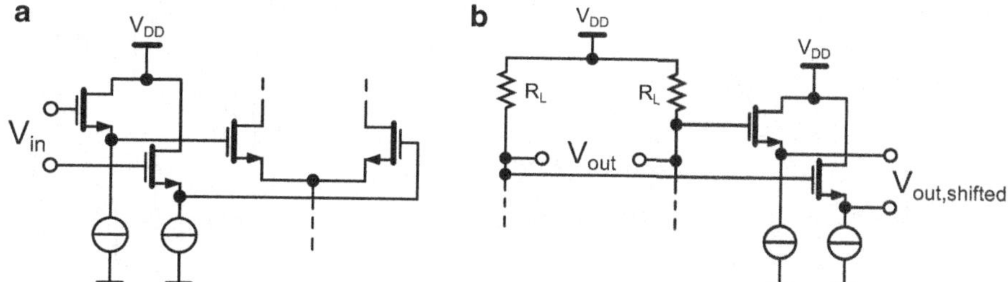

Fig. 2.21 (**a**) Level shifting at the inputs of an MCML gate. (**b**) Level shifting at the output of an MCML gate

- each source–follower stage creates an additional delay, of the same order of magnitude as the gate delay, thus resulting in a consequent speed penalty,
- inserting the level-shifters at the input of the gate results in a very high number of source–followers, resulting in a prohibitive cost in terms of area and power consumption
- inserting the level-shifters at the output of the gates is incompatible with an automated approach, since the load of each output (unshifted and shifted) cannot be determined a-priori, making it impossible to estimate the delay
- source–followers create an important amount of noise during switching

Not shifting the level of the input signals implies allowing the transistors in the differential pair to leave saturation region for a range of input voltages, as given in (2.58). Depending on the value of V_{ov}, this range of voltage may be limited to a small range when $V_1 \approx V_{DD}$, or extend so much as to keep M_1 from saturating most of the time.

One of the consequences of this is that, because the transconductance of the transistors in the differential pair is dependent on their drain voltage when they are not saturated, the transfer characteristic will become asymmetrical unless the drain voltages of both transistors are equal. In the case of the gate of Fig. 2.19, the drain voltages of M_1 and M_2 are very different, resulting in the transfer function displayed in Fig. 2.22a, which plots simulation results for the AND2 network of Fig. 2.19, together with the transfer function of a buffer with identical parameters. As it can be seen, the transfer curve of the asymmetrical network is shifted horizontally, resulting in an important decrease of the noise margin.

Though this decrease can be compensated by an increase in transistor sizes, the asymmetry of the network can also be compensated for, by balancing the network

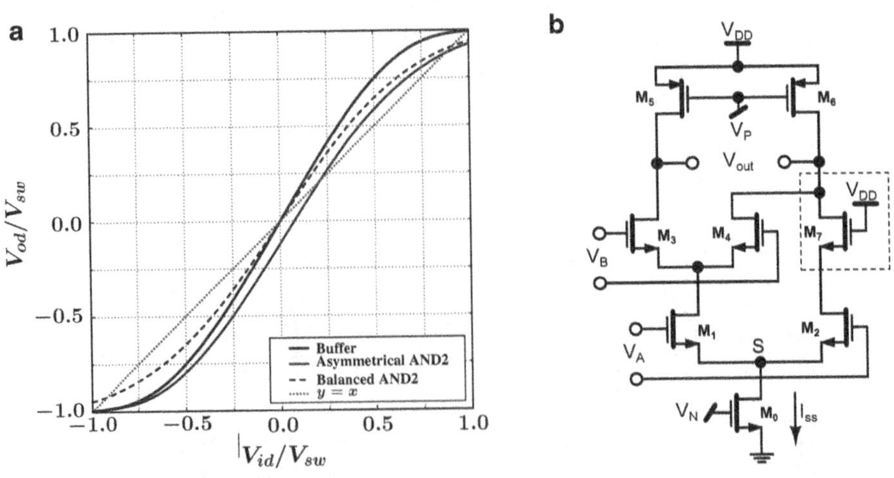

Fig. 2.22 (**a**) Transfer functions of the asymmetric and balanced AND2 logic networks. (**b**) Balancing the AND2 logic network by adding transistor M_7

as depicted on Fig. 2.22b, by adding transistor M_7 to lower the drain voltage of M_2 to a value close to the drain voltage of M_1. In order to achieve balancing, M_7 should have the same size as M_3 and M_4, so that their gate-to-source voltage will be equal at equal drain current. The resulting transfer curve for the balanced network is also displayed in Fig. 2.22a. In this example, the initial noise margin of 100 mV for the buffer drops to 35 mV for the asymmetrical network, while after balancing it restored to 60 mV.

There is another consequence of the transistors not being saturated: when the transistor leaves the saturation region, its transconductance drops—slowly in the beginning, and more sharply when entering deeper in the linear region. If the differential gain of the pair is larger than one when it enters linear region, the noise margin will be affected. This loss in noise margin can be compensated in two ways: by increasing the size of the transistors in the $M_1 - M_2$ differential pair, and by reducing the overdrive voltage of the $M_3 - M_4$ differential pair. In any case, it implies increasing transistor sizes, therefore reducing speed and increasing area.

In Fig. 2.23, simulation results are displayed for a balanced AND2 logic network. These plots show the noise margin of the AND2 gate with respect to input A, which drives the differential pair at the lower level of the network, as a function of transistor sizes. The effect of resizing the $M_1 - M_2$ and $M_3 - M_4$ pairs both independently and simultaneously is considered. As it can be seen, resizing a single pair has a limited effect on the noise margin, which increases with transistor size but quickly saturates. As the plots show, a good compromise is found when increasing both pairs simultaneously. This is confirmed by the simulation results in Fig. 2.24, where the sizes of both differential pairs are swept, and the level curves at constant noise margin are extracted, together with a figure of merit for the area ($A = 2 \cdot W_{12} + 3 \cdot W_{34}$

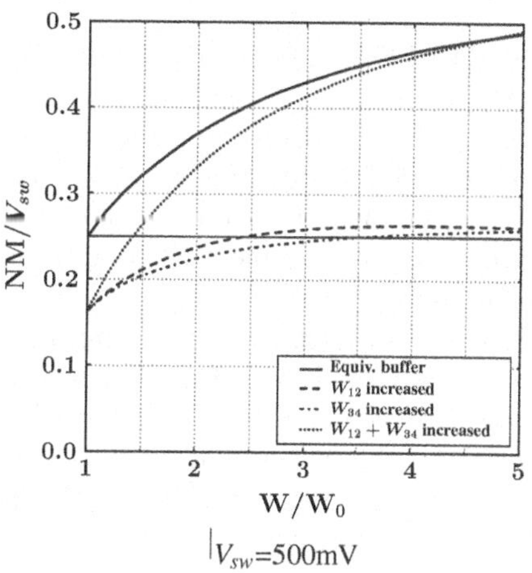

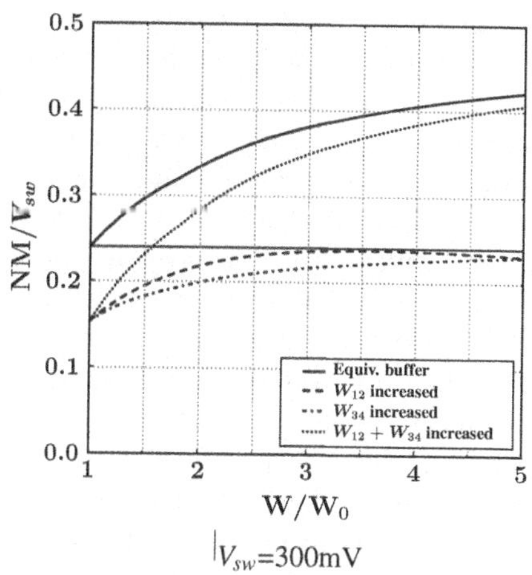

Fig. 2.23 Noise margin compensation in an AND2 logic network without level shifting. The loss in noise margin is compensated by increasing transistor sizes of $M_1 - M_2$ and $M_3 - M_4(-M_7)$ independently and simultaneously

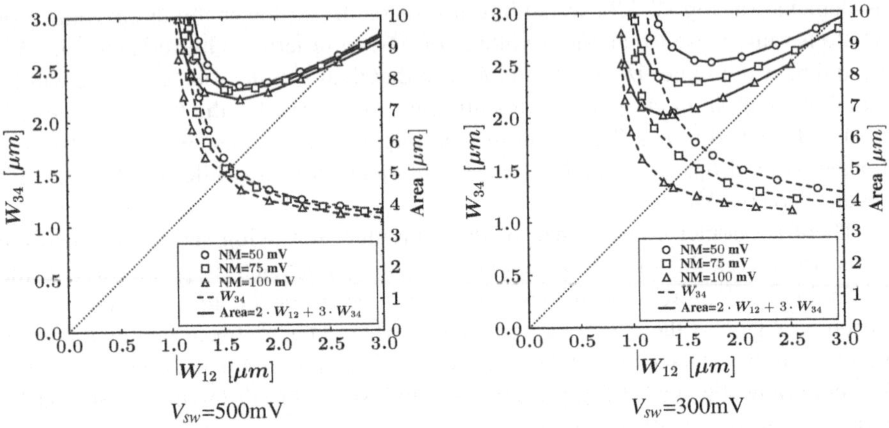

Fig. 2.24 Optimization of AND2 logic network transistor sizes. Simulations were performed with a 90 nm CMOS technology on the balanced AND2 network of Fig. 2.22b

for the AND2 logic network). As predicted, the minimum area is consistently very close to the point where $W_{12} = W_{34}$.

2.4.3 *Dynamic Operation*

The dynamic operation of multi-level MCML logic networks can be analyzed as it was done for the single-stage network in the previous section, by linearizing the differential pair transfer function.

Considering the balanced AND2 logic network of Fig. 2.22b, let us first observe that, when input B is switching, the $M_1 - M_2$ pair is largely unaffected and thus can be approximated by a constant current source. The dynamic behavior is thus exactly the same as for a single-stage network, except for additional parasitics due to increased transistor sizes, and additional connections to the output nodes. The same remains true for any logic network, when considering the switching of an input pair located on the top-most level of the network.

Considering the A input, the behavior is slightly different. In order to linearize the $M_3 - M_4$ differential pair, as well as the M_7 transistor, we observe that their drain current during a switching event is varying between $(1 - \gamma) \cdot I_{SS}$ and $\gamma \cdot I_{SS}$, where γ denotes the effective switched fraction of the tail current, as their gate voltage stays constant and their source voltage is changing. Therefore, they can be approximated by a linear current source of transconductance $g_{ms} = I_{SS}/\Delta V_S$, where ΔV_S denotes the variation in source voltage during a switching event. The variation in source voltage can be calculated from (2.19) as

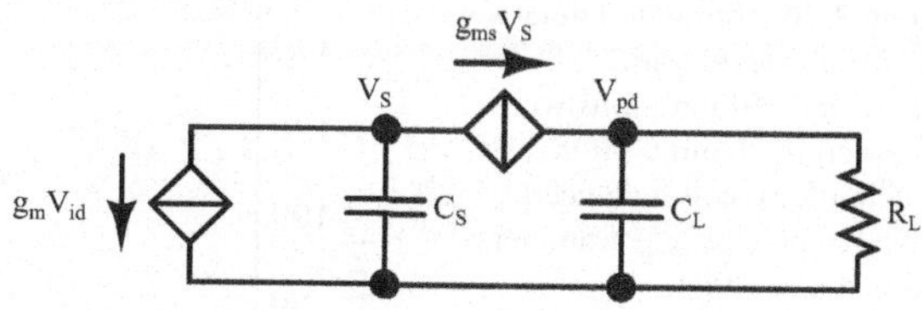

Fig. 2.25 Linear equivalent circuit of a two-level MCML logic network

$$\frac{\Delta V_S}{U_T} = 2\sqrt{2}\sqrt{\frac{I_{SS}}{2I_S}}\left(\sqrt{\gamma} - \sqrt{1-\gamma}\right) + \ln\left(\frac{\gamma}{1-\gamma}\right) \tag{2.59}$$

Substituting (2.42), (2.32), and (2.53), this expression reduces to

$$\Delta V_S = \frac{\alpha}{n} V_{sw} \tag{2.60}$$

suggesting that ΔV_S is close to V_{sw} for practical values of α. This provides a good approximation, therefore we will assume $g_{ms} = 1/R_L$. Writing the differential equations for V_{o1} and V_{o2}, and subtracting we get the differential equation for V_{od}, and the linear equivalent circuit of Fig. 2.25 can be constructed. The transfer function for this circuit is found as

$$\frac{V_{od}}{V_{id}} = \frac{A_V}{(1 + s \cdot \tau_L)(1 + s \cdot \tau_S)} \tag{2.61}$$

where $\tau_L = R_L \cdot C_L$ and $\tau_S = C_S/g_{ms}$. The resulting circuit is of second order, with an additional pole due to the second stage. When $\tau_L \gg \tau_S$, which is true in most cases due to external load capacitance, the second-order system behavior can be approximated by a first-order system with a time constant equal to

$$\tau = \tau_L + \tau_S = (C_L + C_S)\frac{V_{sw}}{I_{SS}} \tag{2.62}$$

The same analysis can be performed for logic networks with n levels, yielding an nth-order system, with a pole for each level. Therefore, the intrinsic propagation delay of a multi-level network will increase for each level of logic, starting from the top, by a constant factor. This is illustrated in Fig. 2.26, where intrinsic delays are plotted for the different inputs of AND gates up to four levels.

2.5 Effect of Nonlinearities

In our previous analyses, we have relied on the linearity of certain elements, and willingly approximated others as linear in order to simplify the calculations. Linearity is a very convenient property, but it is an idealization: in practice, none

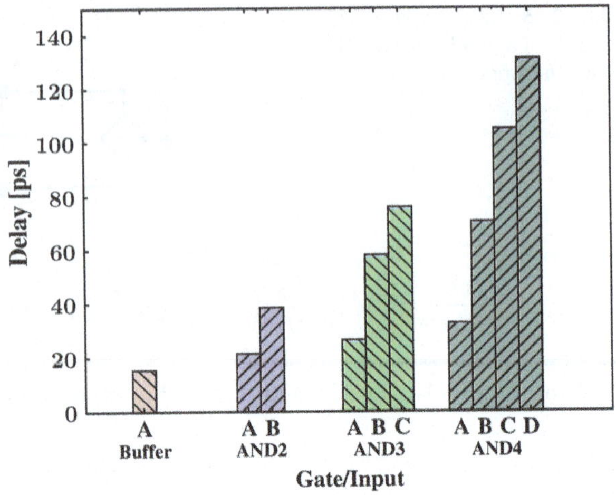

Fig. 2.26 Simulated delays for multi-level logic networks. Simulations were carried in 90 nm CMOS technology with balanced ANDn networks. Parameters are $V_{sw} = 400\,\text{mV}$, $NM = 100\,\text{mV}$

of the real-world devices is linear. Since nonlinearities will cause deviations from the ideal linear behaviors, it is interesting to examine their sources and their effects. Nonlinearities arise from various sources in integrated circuits in general; in MOS current-mode logic circuits in particular, there are three main sources: load devices, differential pairs and junction capacitances.

2.5.1 *Load Devices*

Ideally, load devices should be perfectly linear resistors. In practice, their linearity depends on their implementation: passive components, such as polysilicon resistors, can be highly linear; active devices, on the other hand, are inherently strongly nonlinear and the amount of nonlinearity greatly depends on the biasing and signal swings. In the previous discussions, the load devices have always been approximated by ideal linear resistors. Nonlinearity in the load devices will affect the DC characteristics as well as the transient behavior.

A PMOS device biased in the linear region obeys the following I–V characteristic, as given by the EKV model

$$I_{D,lin} = \beta \cdot n \cdot \left(\frac{V_{DD} - V_G - V_T}{n} - V_{DD} + \frac{V_S + V_D}{2} \right) \cdot (V_D - V_S) \qquad (2.63)$$

A load device has its source connected to the supply voltage and its gate biased at a fixed value, produced by a bias generator circuit to adjust the equivalent resistance. The drain voltage varies between V_{DD} and $V_{DD} - V_{sw}$, which defines the region of interest. In order to approximate the real device by a linear equivalent, one can

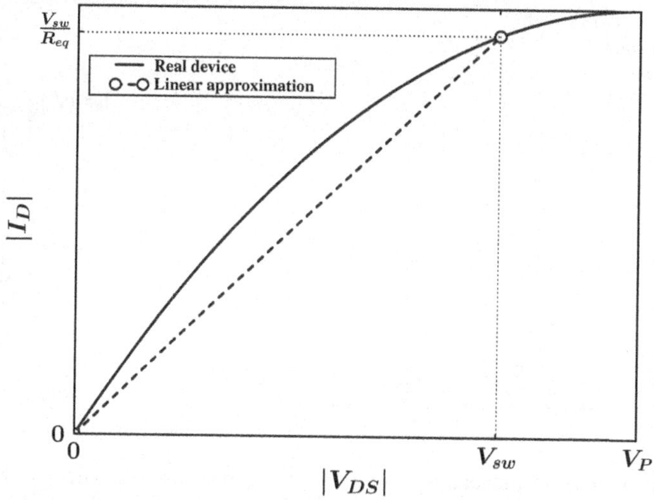

Fig. 2.27 Approximation of nonlinear load device

choose to equate both characteristics at both ends of the region of interest, that is, at $V_{DS} = 0$ and $V_{DS} = -V_{sw}$. With this definition, the equivalent resistance is given by

$$R_{eq} = \frac{V_{sw}}{I_D\left(V_{DS} = -V_{sw}\right)} = \frac{1}{\beta \cdot n \cdot \left(\frac{V_{DD}-V_G-V_T}{n} - \frac{V_{sw}}{2}\right)} \tag{2.64}$$

This definition allows to keep the V_{sw} parameter unchanged. Figure 2.27 illustrates the linear approximation of the load device; as it can be seen, the real resistance is always lower than the equivalent resistance over the region of interest.

The real resistance value is given by $R = |V_{DS}/I_D|$, and the error in resistance due to the linear approximation is

$$\begin{aligned} \Delta R(V_{ds}) &= R_{eq} - R \\ &= \left[\beta \cdot n \cdot \left(\frac{V_{DD} - V_G - V_T}{n} - \frac{1}{2} V_{sw}\right)\right]^{-1} \\ &\quad - \left[\beta \cdot n \cdot \left(\frac{V_{DD} - V_G - V_T}{n} - \frac{1}{2} |V_{DS}|\right)\right]^{-1} \end{aligned} \tag{2.65}$$

The average error over the region of interest is

$$\overline{\Delta R} = \frac{1}{V_{sw}} \int_0^{V_{sw}} \Delta R(V_{DS}) \mathrm{d}V_{DS} = \frac{w}{1 - \frac{1}{2} \cdot w} + \ln\left|1 - \frac{1}{2} \cdot w\right|$$

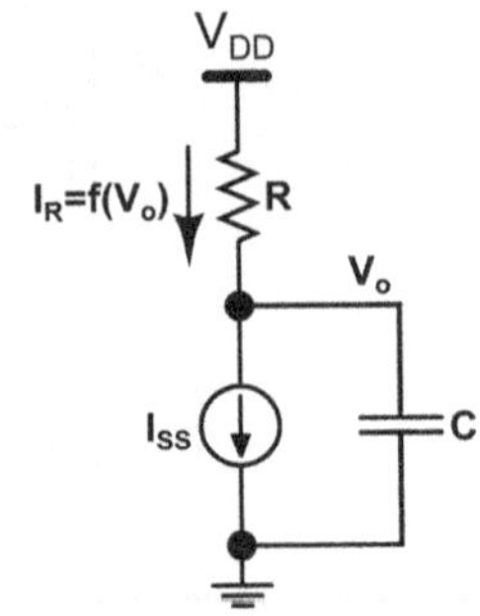

Fig. 2.28 RC network with nonlinear resistance

$$\text{where } w = \frac{n \cdot V_{sw}}{V_{DD} - V_G - V_T} \tag{2.66}$$

Clearly, the error is an increasing function of w. In order to minimize the error, the numerator can be decreased, or the denominator can be increased. In practice, this means using a smaller voltage swing, and using a gate voltage as low as possible, with a supply voltage as high as possible.

The consequences of nonlinearity in the load devices are twofold. First, because the actual resistance is always lower than the approximated linear value, the transfer characteristic of an MCML gate will be flattened, causing a decrease of the noise margin relative to the value calculated with linear load devices. Second, the nonlinearity will cause asymmetrical transient behavior at the rising and falling output nodes, which results in variation of the supply current during switching events. To understand this, let us calculate the step response of a RC network with nonlinear resistance, as shown in Fig. 2.28, where the current through the resistance is voltage-dependent and expressed as

$$\begin{aligned} I_R(V_o) &= \beta \cdot n \cdot \frac{V_{DD} - V_G - V_T}{n} \cdot (V_{DD} - V_o) - \frac{1}{2} \cdot \beta \cdot n \cdot (V_{DD} - V_o)^2 \\ &= A \cdot (V_{DD} - V_o) - \frac{A}{2B} \cdot (V_{DD} - V_o)^2 \end{aligned} \tag{2.67}$$

Such a network obeys the differential equation

$$C_L \frac{\mathrm{d}}{\mathrm{d}t}(V_{DD} - V_o) = I_{SS} - A \cdot (V_{DD} - V_o) + \frac{A}{2B} \cdot (V_{DD} - V_o)^2 \tag{2.68}$$

which is a particular instance of the Riccatti equation, with constant coefficients. Solving the differential equation with the initial condition for the rising and falling transitions, we get

$$V_{o(rise)} = V_{DD} - V_{sw} \cdot \frac{1 - \tanh(t/\tau_r)}{1 + \left(1 - \frac{V_{sw}}{V_P}\right) \cdot \tanh(t/\tau_r)} \tag{2.69}$$

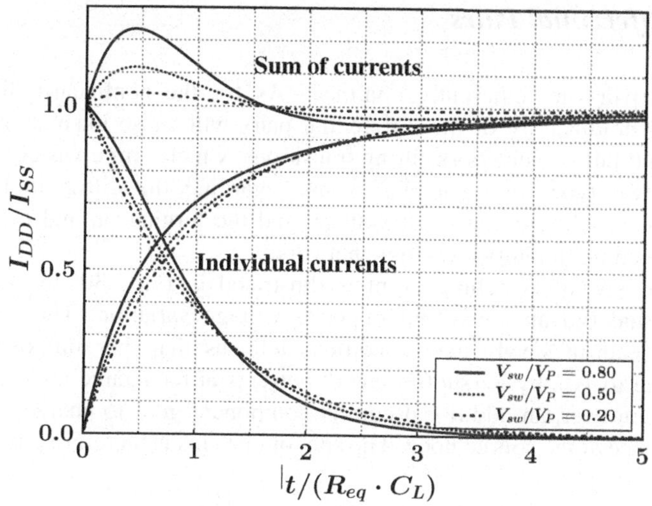

Fig. 2.29 Supply noise caused by the nonlinearity of the load devices

$$V_{o(fall)} = V_{DD} - V_P \frac{1 - \left(1 - \frac{V_{sw}}{V_P}\right)^2}{1 + \left(1 - \frac{V_{sw}}{V_P}\right) \cdot \coth\left(t/\tau_f\right)} \quad (2.70)$$

where

$$V_P = \frac{V_{DD} - V_G - V_T}{n}$$

$$\frac{\tau_r}{R_{eq} \cdot C_L} = \frac{V_P}{V_{sw}} \left[1 - \left(1 - \frac{V_{sw}}{V_P}\right)^2\right]$$

$$\tau_f = \tau_r / \left(1 - \frac{V_{sw}}{V_P}\right)$$

The resulting values of $V_{o(rise)}$ and $V_{o(fall)}$ can be inserted back into (2.67) and added to obtain the total current flowing to the supply node as a function of time. Clearly, the resulting current is not a constant, and therefore an amount of current noise is generated due to the nonlinearity of the load devices. The shape of the current waveform depends on the ratio V_{sw}/V_P, and sample curves are plotted in Fig. 2.29. As it can be guessed, the supply noise increases with increasing V_{sw}/V_P ratio, i.e. as the load devices become more nonlinear. For a typical value of V_{sw}/V_P of 1/2, the amplitude of the current spike is about 10% of I_{SS}.

2.5.2 Differential Pairs

Differential pairs are inherently nonlinear. As for the nonlinearity in the load devices, the nonlinearity of the differential pairs will cause mismatches between the rising and falling behaviors during transitions, which sum up as current spikes. There are two types of switching events regarding the differential pairs: the switching of the differential input voltage, and the turning on and off of the tail current happening in multi-level logic networks.

The first type of switching event is illustrated in Fig. 2.30: the gates of the transistors undergo an equal and opposite voltage variation. The study of the differential pair in steady-state conditions tells us that the sum of currents is always equal to I_{SS}, by Kirshoff's law of currents at the source node. In transient conditions however, an additional current component goes to charge or discharge the capacitance at the source node. The sum of currents is therefore equal to

$$I_1 + I_2 = I_{SS} + C_S \frac{\mathrm{d}V_S}{\mathrm{d}t} \tag{2.71}$$

Note that, if the transistors behaved perfectly linearly with respect to their gate voltage, the source voltage V_S would be constant during a transition. However, as it was shown earlier, the source voltage varies during transitions as given by formula (2.60). The variation of the source voltage can be written as

$$\frac{\mathrm{d}V_S}{\mathrm{d}t} = \frac{\mathrm{d}V_S}{\mathrm{d}V_{id}} \cdot \frac{\mathrm{d}V_{id}}{\mathrm{d}t} \tag{2.72}$$

Therefore, the peak of current strongly depends on the rise time of the input signal. Essentially, this noise component will grow with the voltage swing, the capacitance at the node S (which consists of a constant part and a part proportional to transistor sizes, due to their source diffusion), and the speed of the input signal.

The second type of switching event in a differential pair is when the differential input voltage is constant and the tail current is switched on or off. This happens in gates with multiple levels of stacked differential pair, when a pair at a lower level is switching. Typically, such a switching event will involve simultaneously turning one pair on and another one off; because of the nonlinear behavior of the devices,

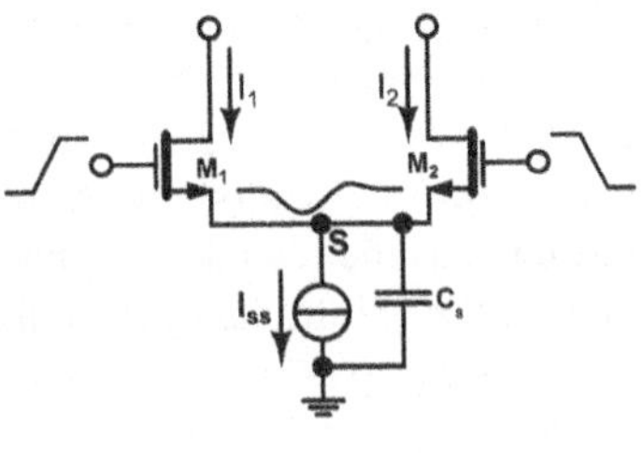

Fig. 2.30 Noise generated by the switching of a differential pair input

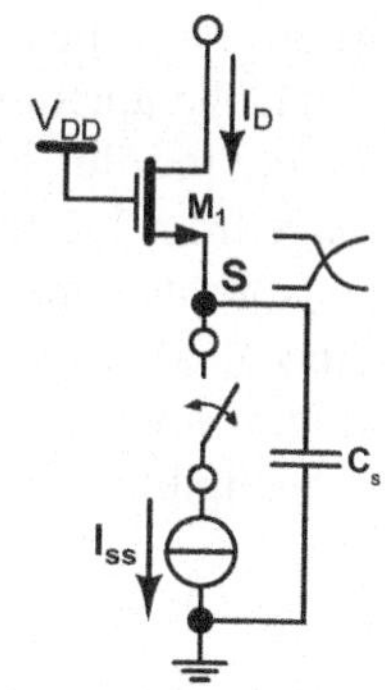

Fig. 2.31 Noise generation during the turning on/off of a differential pair tail current

the two current waveforms will not compensate each other and sum up as current spikes. This type of switching event is illustrated in Fig. 2.31.

In this process, if we assume that the switching is instantaneous and reduce the transistor model to its strong inversion behavior, we can write the differential equation at node S as

$$I_D = \frac{n \cdot \beta}{2}(V_P - V_S)^2 = C_S \cdot \frac{\mathrm{d}V_S}{\mathrm{d}t} + I_{SS} \tag{2.73}$$

The solution of this differential equation is different when the pair is turning on and when it is turning off,

$$I_D(t) = \begin{cases} \dfrac{I_{SS}}{\left(1+\frac{I_{SS}}{C_S}\sqrt{\frac{n\cdot\beta}{2I_{SS}}}\cdot t\right)^2} & \text{when turning OFF} \\ I_{SS}\cdot\tanh^2\left(\frac{I_{SS}}{C_S}\sqrt{\frac{n\cdot\beta}{2I_{SS}}}\cdot t\right) & \text{when turning ON} \end{cases}$$

Clearly, the sum of those two currents cannot be null, and the resulting spikes depend on the value of C_S here also.

2.5.3 *Junction Capacitances*

The capacitance of the p–n junctions that form the drains and sources of the transistors typically accounts for a significant part of the internal parasitics. The capacitance of a p–n junction is nonlinear because the depletion region width changes with the bias voltage. An expression for the C–V relationship is found in [9]

$$C_j = \frac{C_{j0}}{\left(1+\frac{V_R}{V_{bi}}\right)^m} \tag{2.74}$$

where V_R is the (reverse) bias voltage across the junction, V_{bi} is the built-in voltage, C_{j0} is the junction capacitance at zero bias, and m is an exponent depending on the doping profile. Typical values of V_{bi} are in the range of 0.5–1 V, and values of m range from $1/3$ to $1/2$.

Note that the capacitance becomes more linear as the bias voltage increases. Since in MCML circuits the voltage swings are small and all nodes have a DC bias of at least several hundreds of millivolts, the effect of nonlinear junction capacitances is negligible for all practical purposes.

2.5.4 *Overall Noise Performance*

The different nonlinear effects that have been discussed separately are in practice all happening together, and affecting each other. It is hardly possible to predict accurately the result of simultaneous effects; some might tend to add up, while others might tend to cancel. It is thus interesting to construct an overall picture of the noise performance of MCML gates by simulations, with respect to the various design parameters.

Of all the contributions, the load device nonlinearity is probably the most important, because all other contributions depend on internal capacitances or time constants, and are filtered by the external load before reaching the supply. Since the load at the output nodes is typically much larger than at any internal node, all the internal current spikes tend to vanish at least partly in real situations. The spike of current due to the nonlinear loads, on the other hand, has a constant amplitude regardless of the switching speed of the output nodes.

Since as we have seen, the phenomena that cause the current spikes are transient, and depend on the signal rise times and shape, it is not meaningful to test a circuit in an ideal environment. In order to obtain pertinent results, a common test setup is to construct a ring oscillator out of the device to be tested. In this way, the device is functioning at a speed which is realistic, and all the signals in the circuit are real gate outputs and thus have a realistic shape.

Figure 2.32 plots the peak-to-peak current ripple on the supply obtained by simulation of ring oscillators for different gates sized for different voltage swings and loading conditions. As it can be seen, the global noise decreases when the voltage swing is increased. This is due to the decrease of internal parasitics due to the smaller transistor sizes. Also, it is clear that the noise produced by multi-level logic networks is more important, especially during the switching of differential pairs located at the lower levels of the network.

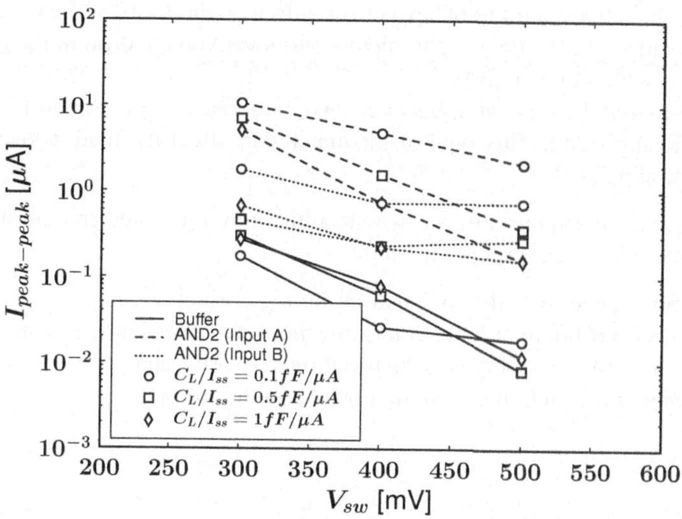

Fig. 2.32 Supply noise in an MCML ring oscillator as a function of the design parameters. Simulation carried with 90 nm CMOS technology, at $I_{SS} = 100\,\mu$A and $NM = 100$ mV

2.6 Random Effects

Environmental variations will inevitably affect the operation of MCML circuits. These include variations in the device characteristics, due to process variations and mismatch, as well as operating supply voltage and temperature variations. MCML circuits are rather tolerant to environmental fluctuations since key performance parameters—speed and power dissipation of individual gates—are adjusted through bias voltages. An on-chip bias generation circuit should provide these voltages to ideally keep the voltage swing and tail current well-defined and constant. Though, strictly speaking, it is not necessary to use a bias generator, it is frequently used to counterbalance process variations.

Variations may appear at different levels, causing deviations from the expected characteristics:

- **Process and environmental variations:** variations affecting all devices equally. These variations include process variations due to the manufacturing, global supply voltage variations, and global temperature variations. In a closed loop approach involving a bias generator circuit, global variations are cancelled by adjusting the different parameters to the external references. However, these references may have a certain amount of uncertainty. In addition, imperfections in the bias generator (or absence thereof) will cause the adjusted parameters to vary.
- **On-chip variations:** environmental differences between the reference circuit used to generate the bias voltages and the circuit receiving the bias voltages,

causing the latter circuit to behave differently than the former. These include on-chip variations of the device parameters and local voltage drop in the power and ground distribution networks.

- **Local mismatch:** mismatch between two components assumed to be identical, within a single cell. This type of mismatch can affect the load devices and the differential pairs.

In the study of random effects, we are ultimately interested in determining the resulting variation in several parameters:

- **Operating speed**, in order to construct timing-correct circuits,
- **Power dissipation**, in order to determine the worst-case power consumption, and appropriately dimension power distribution networks, and
- **Noise margin**, which must remain positive in the worst-case.

2.6.1 Process Variations

Process variations, also called chip-to-chip variations, describe the fluctuation of the mean value of device characteristics. This kind of variation globally affects all devices on the chip.

As it has been stated, the use of a regulation circuit (bias generator) is frequent to cancel the process variations. In the absence of some form of regulation, it is hardly possible to keep the voltage swing under control, within a safe operating range.

Typically, a replica scheme is used, as illustrated in Fig. 2.33, where a circuit block receives bias voltages from an on-chip bias control block, which generates the bias voltages based on a reference voltage and a reference current. The bias

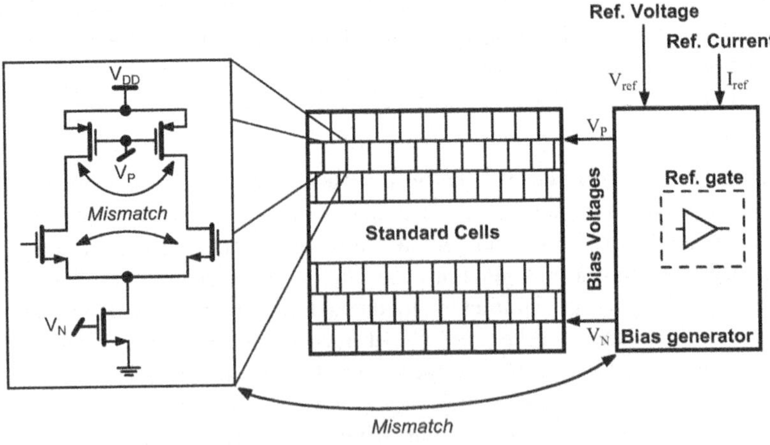

Fig. 2.33 MCML circuit with biasing and on-chip parameter variations

Table 2.1 Summary of sensitivities to process variations

	I_{SS}	V_{sw}	t_d	P	NM
V_{TN}	−	−	−	−	−
β_N	+	+	+	+	+
V_{TP}		+	+		+
β_P		−	−		−
V_{DD}		−	−	+	−

voltages are adjusted to tune the performance parameters of a reference gate to the desired values. The generated bias voltages are distributed to all the standard cells in the circuit, in order to match their parameters to the parameters of the reference gate.

Implementing such a closed-loop regulation allows to cancel process variations nearly perfectly, with the drawback that it necessitates external current and voltage references. When using simpler schemes, that cancel process variations to a limited extent, or when using no regulation at all, a classical corner analysis approach can be adopted. In such an approach, the circuit performance is evaluated for extreme (maximum and minimum) values of the process parameters, in order to obtain best-case and worst-case performance.

Typically, two corner cases are defined for each type of device: fast and slow. The fast corner is when β is large and V_T is small, and vice versa. Table 2.1 summarizes the sensitivities of the main performance parameters of MCML circuits to the various device and environmental parameters, in the case where no regulation is used. The sign of the sensitivity denotes whether a variation in a parameter is causing a variation of the same (+) or opposite (–) sign of the corresponding quantity. As this table shows, the worst case for noise margin is the slow-NMOS, fast-PMOS, high-V_{DD} case. Regarding speed, the worst case is the fast-NMOS, slow-PMOS, low-V_{DD} case. For power, the worst case is the fast-NMOS, high-V_{DD} case and is independent of the characteristics of PMOS devices. It is to be noted that these sensitivities can be different when using on-chip regulation, depending on the specific implementation of the bias generator.

2.6.2 *On-Chip Variations and Mismatch*

On-chip variations (OCV) describe the fluctuations of parameters around their mean value, for devices on the same chip. The consequence of these variations is that different circuits on the same chip, with the same nominal parameters, will behave differently. In particular, when a closed-loop approach is adopted to cancel process variations, mismatch in parameters between the devices in the bias generator and other devices in the circuit will cause the speed, power, and noise margin to vary across the chip. In addition to pure device parameters, local voltage drops on the power and ground distribution networks and local heating will cause additional variations.

Classically, on-chip variations are described statistically by gaussian distributions. Each parameter P_i is characterized by its standard deviation $\sigma(P_i)$. The classical model for MOSFET parameter variation expresses the standard deviation of the difference in threshold voltage and current factor between two devices of channel dimensions W, L and separated by a distance D as [13]

$$\sigma^2(\Delta V_T) = \frac{A_{V_T}^2}{W \cdot L} + S_{V_T}^2 \cdot D^2 \tag{2.75}$$

$$\frac{\sigma^2(\Delta\beta)}{\beta^2} = \frac{A_\beta^2}{W \cdot L} + S_\beta^2 \cdot D^2 \tag{2.76}$$

where A is the area proportionality constant, and S the spacing proportionality constant. The influence of spacing is typically very small and can be neglected [16]. Note that this describes the *difference* of parameter values between two identical devices; the parameter variations for individual devices are thus given by

$$\sigma(P_i) = \frac{\sigma(\Delta P_i)}{\sqrt{2}} \tag{2.77}$$

Provided that the variations are small enough so that a quantity $Q(P_1, \ldots, P_i, \ldots, P_N)$ can be well approximated by linearization around the nominal value of the parameters, the standard deviation in Q caused by variations in P_i can be approximated by

$$\sigma_{P_i}(Q) \approx \frac{\partial Q}{\partial P_i} \cdot \sigma(P_i) \tag{2.78}$$

where $\partial Q/\partial P_i$ is sensitivity of Q to P_i The total variation in Q caused by the simultaneous random variations of all N parameters $P_1, \ldots, P_N$, provided they are statistically independent, is given by

$$\sigma(Q) \approx \sqrt{\sigma_{P_1}^2 + \cdots + \sigma_{P_i}^2 + \cdots + \sigma_{P_N}^2} \tag{2.79}$$

Random parameters β and V_T can be accounted for using this statistical approach. Using the resulting value of $\sigma(Q)$, a worst-case value can be chosen, typically as $\pm 3\sigma$. However, other parameters like the local voltage drop on the supply can hardly be characterized by a statistical distribution; for these parameters, an additional margin can be added to the worst-case statistical value, that can be calculated using the derivatives as

$$\max(\Delta Q) \approx \left|\frac{\partial Q}{\partial V_{DD}}\right| \max(\Delta V_{DD}) + \left|\frac{\partial Q}{\partial V_{GND}}\right| \max(\Delta V_{GND}) \tag{2.80}$$

Table 2.2 Sensitivities of I_{SS}

Device	Parameter	Sensitivity
Tail current source	$V_{TN}, -V_N, V_{GND}$	$\frac{\sigma(I_{SS})}{I_{SS}} = 2\frac{\sigma(V_{TN}, V_N, V_{GND})}{V_N - V_{TN}}$
Tail current source	β_N	$\frac{\sigma(I_{SS})}{I_{SS}} = \frac{\sigma(\beta_N)}{\beta_N}$

Table 2.3 Sensitivities of V_{sw}

Device	Parameter	Sensitivity
PMOS load device	$V_{TP}, V_P, -V_{DD}$	$\frac{\sigma(V_{sw})}{V_{sw}} = \frac{\sigma(V_{TP}, V_P, V_{DD})}{V_{DD} - V_P - V_{TP} - nV_{sw}}$
PMOS load device	β_P	$\frac{\sigma(V_{sw})}{V_{sw}} = -\frac{V_{DD} - V_P - V_{TP} - \frac{n}{2}V_{sw}}{V_{DD} - V_P - V_{TP} - nV_{sw}} \cdot \frac{\sigma(\beta_P)}{\beta_P}$
Tail current source	I_{SS}	$\frac{\sigma(V_{sw})}{V_{sw}} = \frac{V_{DD} - V_P - V_{TP} - \frac{n}{2}V_{sw}}{V_{DD} - V_P - V_{TP} - nV_{sw}} \cdot \frac{\sigma(I_{SS})}{I_{SS}}$

Tail Current and Voltage Swing Using the expressions developed in the previous sections, we can calculate the expressions for the sensitivities of the tail current and voltage swing with respect to these various parameters. These expressions are summarized in Tables 2.2 and 2.3.

Note that, in the case where a regulation scheme is used, the parameter variations $\sigma(\Delta P_i)$ describe the *differences* of a device's parameter values with respect to the reference device. If no regulation is used, these variations should be replaced by that of individual devices.

Note also that, in the case of the voltage swing, the presence of on-chip variations will cause the complementary outputs of each gate to display two different, independent voltage swings. This is important to consider when analyzing the effect of variations on the delay and noise margin.

Delay To calculate the variation in the delay, we should consider the asymmetry in the voltage swings caused by on-chip variations. This results in two different time constants for the two output nodes, which can be expressed by

$$\frac{\Delta\tau}{\tau} = \frac{\Delta R_L}{R_L} \tag{2.81}$$

where

$$\tau = \frac{\tau_1 + \tau_2}{2} \quad \text{and} \quad \Delta\tau = \tau_1 - \tau_2 \tag{2.82}$$

The step response given by (2.47) in the absence of mismatch is now given by

$$\frac{V_{o,d}}{V_{sw}} = \left(1 - \frac{1}{2}\frac{\Delta R_L}{R_L}\right) \cdot \left[1 - e^{-\frac{t}{\tau_2}}\right] - \left(1 + \frac{1}{2}\frac{\Delta R_L}{R_L}\right) \cdot e^{-\frac{t}{\tau_1}} \tag{2.83}$$

With the definitions of τ and $\Delta\tau$, we can write

$$\frac{t}{\tau_{1,2}} = \frac{t}{\tau \pm \Delta\tau} = t \cdot \frac{\tau \mp \Delta\tau}{\tau^2 - (\Delta\tau)^2} \approx \frac{t}{\tau} \cdot \left(1 \mp \frac{\Delta\tau}{\tau}\right) \tag{2.84}$$

where it was assumed that $\Delta\tau/\tau$ is small enough so that $(\Delta\tau/\tau)^2 \approx 0$. With this simplification, (2.83) can be rewritten as

$$\frac{V_{o,d}}{V_{sw}} \approx 1 - 2\mathrm{e}^{-\frac{t}{\tau}} \cdot \cosh\left(\frac{1}{2}\frac{t}{\tau}\frac{\Delta\tau}{\tau}\right) \pm \frac{1}{2}\frac{\Delta\tau}{\tau}\left[1 - \mathrm{e}^{-\frac{t}{\tau}} \cdot \sinh\left(\frac{1}{2}\frac{t}{\tau}\frac{\Delta\tau}{\tau}\right)\right] \tag{2.85}$$

which can be further approximated as

$$\pm\frac{V_{o,d}}{V_{sw}} \approx 1 - 2\mathrm{e}^{-\frac{t}{\tau}} \pm \frac{1}{2}\frac{\Delta\tau}{\tau} \tag{2.86}$$

resulting in a delay of

$$t_d \approx -\tau \cdot \ln\left(\frac{1}{2} \pm \frac{1}{4}\frac{\Delta\tau}{\tau}\right) \approx \tau \cdot \ln(2) \pm \frac{1}{2}\Delta\tau \tag{2.87}$$

where the $\pm$ sign indicates that there are different delays for the rising and falling transitions. Note that there are two components, one depending on the average value of the time constants, and one depending on their difference.

The variations for each individual time constant can be calculated from the sensitivities summarized in Table 2.4. In order to calculate the statistics of their average value $\tau = (\tau_1 + \tau_2)/2$ and difference $\Delta\tau = \tau_1 - \tau_2$, we must not overlook the fact that, though variations in the load device parameters are independent for the two output nodes, the variation in the tail current affects both time constants in the same way. Therefore, the tail current does not contribute to the difference of time constants.

$$\sigma^2(\Delta\tau) = \sigma^2(\tau_1 - \tau_2) \approx 2\left(\frac{\partial\tau_{1,2}}{\partial V_{TP}}\right)^2 \sigma^2(V_{TP}) + 2\left(\frac{\partial\tau_{1,2}}{\partial\beta_P}\right)^2 \sigma^2(\beta_P) \tag{2.88}$$

Statistically speaking, this is expressed as a *correlation* between the two values. The statistical correlation factor ρ can be calculated as

$$\rho_\tau = 1 - \frac{\sigma^2(\Delta\tau)}{2\sigma^2(\tau)} \tag{2.89}$$

Table 2.4 Sensitivities of $\tau_{1,2}$

Device	Parameter	Sensitivity
PMOS load device	$V_{TP}, V_P, -V_{DD}$	$\frac{\sigma(\tau_{1,2})}{\tau} = \frac{\sigma(V_{TP}, V_P, V_{DD})}{V_{DD} - V_P - V_{TP} - nV_{sw}}$
PMOS load device	β_P	$\frac{\sigma(\tau_{1,2})}{t_d} = -\frac{V_{DD} - V_P - V_{TP} - \frac{n}{2}V_{sw}}{V_{DD} - V_P - V_{TP} - nV_{sw}} \cdot \frac{\sigma(\beta_P)}{\beta_P}$
Tail current source	I_{SS}	$\frac{\sigma(\tau_{1,2})}{t_d} = \frac{1}{2} \frac{nV_{sw}}{V_{DD} - V_P - V_{TP} - nV_{sw}} \cdot \frac{\sigma(I_{SS})}{I_{SS}}$

Table 2.5 Sensitivities of t_d

Device	Parameter	Sensitivity
Multiple devices	τ	$\frac{\sigma(t_d)}{t_d} = \frac{\sigma(\tau)}{\tau}$
Multiple devices	$\Delta\tau$	$\frac{\sigma(t_d)}{t_d} = -\frac{\sigma(\Delta\tau)}{2\tau \ln(2)}$

so that the variance of the average time constant is given by

$$\sigma(\tau) = \sqrt{2}\sigma(\tau_{1,2})\sqrt{1 + \rho_\tau} \tag{2.90}$$

The sensitivities of the delay t_d are summarized in Table 2.5.

Noise Margin To assess the effect of parameter variation on the noise margin, we must again consider the effect of asymmetry. The differential pair transfer characteristic (2.21) can be written, in the presence of mismatch, as

$$\frac{V_{id}}{2nU_T} = \frac{\Delta V_{TN}}{2nU_T} + \sqrt{\frac{I_{SS}}{2I_S}} \cdot \left(\sqrt{\frac{1 + \frac{\Delta I}{I_{SS}}}{1 + \frac{\Delta \beta_N}{2\beta_N}}} - \sqrt{\frac{1 - \frac{\Delta I}{I_{SS}}}{1 - \frac{\Delta \beta_N}{2\beta_N}}} \right) + \tanh^{-1}\left(\frac{\Delta I}{I_{SS}}\right) - \tanh^{-1}\left(\frac{\Delta\beta}{2\beta}\right) \tag{2.91}$$

The variation in noise margin due to variations in the load device due to V_{TP} and β_P, tail current I_{SS} and average current factor β_N of the differential pair transistors can be deduced by obtaining the sensitivities from (2.21). However, as suggested by (2.91), additional variations are due to the mismatch in threshold voltage and current factors in the differential pair. To a first approximation, this results in an horizontal shift of the transfer characteristic, which directly subtracts to the noise margin. The amount of shifting can be found by setting $\Delta I = 0$ in (2.91) and solving for V_{id}, resulting in

$$\Delta NM \approx \Delta V_T - nU_T \cdot \left[1 + \sqrt{\frac{I_{SS}}{2I_S}} \right] \cdot \frac{\Delta\beta}{\beta} \tag{2.92}$$

The resulting sensitivities are summarized in Table 2.6.

Table 2.6 Sensitivities of $NM_{H,L}$

Device	Parameter	Sensitivity
PMOS load device	$V_{TP}, V_P, -V_{DD}$	$\frac{\sigma(NM_{H,L})}{\sqrt{1-1/\xi}} = V_{sw} \cdot \frac{\sigma(V_{TP}, V_P, V_{DD})}{V_{DD} - V_P - V_{TP} - nV_{sw}}$
PMOS load device	β_P	$\frac{\sigma(NM_{H,L})}{\sqrt{1-1/\xi}} = V_{sw} \cdot \frac{V_{DD} - V_P - V_{TP} - \frac{n}{2}V_{sw}}{V_{DD} - V_P - V_{TP} - nV_{sw}} \cdot \frac{\sigma(\beta_P)}{\beta_P}$
Tail current source	I_{SS}	$\frac{\sigma(NM_{H,L})}{\sqrt{1-1/\xi}} = \left[V_{sw} \cdot \frac{V_{DD} - V_P - V_{TP} - \frac{n}{2}V_{sw}}{V_{DD} - V_P - V_{TP} - nV_{sw}} - nU_T\sqrt{\frac{I_{SS}}{2I_S}}\right] \frac{\sigma(I_{SS})}{I_{SS}}$
Differential pair	β_N	$\frac{\sigma(NM_{H,L})}{\sqrt{1-1/\xi}} = nU_T\sqrt{\frac{I_{SS}}{2I_S}} \cdot \frac{\sigma(\beta_N)}{\beta_N}$
Differential pair	$\Delta\beta_N$	$\sigma(NM_{H,L}) = -nU_T\left[1 + \sqrt{\frac{I_{SS}}{2I_S}}\right] \cdot \frac{\sigma(\Delta\beta_N)}{\beta_N}$
Differential pair	ΔV_{TN}	$\sigma(NM_{H,L}) = \sigma(\Delta V_{TN})$

We must note here also that the two noise margin NM_H and NM_L are not statistically independent. Variations in load device parameters are independent, whereas variations in the tail current I_{SS} and the average current factor β_N of the differential pair affect both noise margins equally. Moreover, variations due to mismatch in the current factors and threshold voltage in the differential pair affect both noise margin opposedly. Therefore, the difference of the noise margins is characterized by

$$\begin{aligned}\sigma^2(\Delta NM) &= \sigma^2(NM_H - NM_L)\\ &\approx 2\left(\frac{\partial NM_{H,L}}{\partial V_{TP}}\right)^2 \sigma^2(V_{TP}) + 2\left(\frac{\partial NM_{H,L}}{\partial \beta_P}\right)^2 \sigma^2(\beta_P)\\ &\quad +4\cdot\left(\frac{\partial NM_{H,L}}{\partial \Delta\beta_N}\right)^2 \sigma^2(\Delta\beta_N) + 4\cdot\left(\frac{\partial NM_{H,L}}{\partial \Delta V_{TN}}\right)^2 \sigma^2(\Delta V_{TN})\end{aligned} \tag{2.93}$$

and there is a correlation factor given by

$$\rho_{NM} = \frac{\sigma^2(\Delta NM)}{2\sigma^2(NM)} \tag{2.94}$$

The effective noise margin is defined as $NM = \min(NM_H, NM_L)$. An expression for the mean and variance of the minimum of two gaussian random variables is found in [12] as

$$\mu(NM) = \mu\left[\min(NM_H, NM_L)\right] = \mu(NM_{H,L}) - \sigma(NM_{H,L})\sqrt{\frac{1-\rho_{NM}}{\pi}} \tag{2.95}$$

$$\sigma(NM) = \sigma\left[\min(NM_H, NM_L)\right] = \sigma(NM_{H,L})\sqrt{1 - \frac{1-\rho_{NM}}{\pi}} \tag{2.96}$$

2.6.3 Numerical Example

In order to verify the models presented previously, and to get a representative idea of the magnitude of random effects, a numerical example is presented hereafter. In this example, a simple MCML buffer is analyzed with the technological data from a 90 nm CMOS process. The gate is designed for a nominal supply voltage of $V_{DD} = 1\,\text{V}$, tail current of $I_{SS} = 30\,\mu\text{A}$, voltage swing of $V_{sw} = 400\,\text{mV}$, and noise margin of $NM = 100\,\text{mV}$. The process mismatch data gives A_{V_T} and A_β allowing to calculate the variations in threshold voltage and current factor for each transistor as follows:

$$\sigma(\Delta P) = \frac{A_P}{\sqrt{W \cdot L}}$$

$$\sigma(P) = \frac{\sigma(\Delta P)}{\sqrt{2}}$$

The resulting transistor sizes and parameter variations are summarized in Table 2.7.

Using a replica bias generator, the circuit is simulated across process corners. The results are given in Table 2.8.

Monte-Carlo sampling is then performed to simulate the effect of parameter variations. The results are summarized in Table 2.9, with the calculated numerical values for the sake of comparison. The simulation was performed at nominal process corner, and the calculations performed with the values of V_T and β obtained from the DC operating point. Note that the replica bias circuit generates V_N as

$$V_N = V_{TN} + \sqrt{\frac{2nI_{SS}}{\beta_N}} \tag{2.97}$$

Table 2.7 MCML buffer transistor sizes and parameter variation

Device	Type	Width	Length	$\sigma(V_T)$	$\sigma(\beta)/\beta$
PMOS load device	LVT[a]	0.42 μm	0.24 μm	6.26 mV	1.87%
Tail current source	LVT[a]	0.42 μm	0.24 μm	10.5 mV	2.78%
Differential pair	RVT[b]	0.48 μm	0.08 μm	11.3 mV	3.52%

[a] Low-V_T
[b] Regular-V_T

Table 2.8 MCML buffer parameters across process corners

Parameter	Corner				
	TT[a]	SS[b]	FF[c]	SNFP[d]	FNSP[e]
V_N [mV]	498	597	416	534	464
V_P [mV]	285	18	509	336	226
I_{SS} [μA]	30	30	30	30	30
V_{sw} [mV]	407	411	405	408	406
NM [mV]	111	98	126	109	113
t_d [ps] ($C_L = 5$ fF)	58.2	62.5	54.1	57.9	58.1
P [μW]	30.1	27.1	33.0	30.0	30.1

[a] Typical NMOS – typical PMOS – 27 °C – $V_{DD} = 1.0$ V
[b] Slow NMOS – slow PMOS – 85 °C – $V_{DD} = 0.9$ V
[c] Fast NMOS – fast PMOS – 40 °C – $V_{DD} = 1.1$ V
[d] Slow NMOS – fast PMOS – 27 °C – $V_{DD} = 1.0$ V
[e] Fast NMOS – slow PMOS – 27 °C – $V_{DD} = 1.0$ V

Table 2.9 Numerical results for OCV

Parameter	Calculated	Monte-Carlo
$\sigma(V_N)$ [mV]	4.03	4.62
$\sigma(V_P)$ [mV]	3.55	4.45
$\sigma(I_{SS})$ [μA]	2.27	2.06
$\sigma(V_{sw})$ [mV]	60.3	60.5
$\sigma(\tau)$ [ps]	6.93	5.60
$\sigma(\Delta\tau)$ [ps]	5.74	5.28
ρ_τ [–]	0.71	
$\sigma(t_d)$ [ps]	5.60	3.88
$\sigma(NM_{H,L})$ [mV]	33.6	38.1
$\mu(\overline{NM})$ [mV]	91.3	98.2
$\sigma(\overline{NM})$ [mV]	35.3	32.2
$\sigma(\Delta NM)$ [mV]	40.2	40.5
ρ_{NM} [–]	0.28	
$\sigma(NM)$ [mV]	35.3	33.6

and V_P as

$$V_P = V_{TN} + \sqrt{\frac{2nI_{SS}}{\beta_N}} \tag{2.98}$$

The sensitivities are deduced from these equations. Furthermore, devices are replicated eight times in parallel in the bias generator, so that their variations average out, reducing the spread of V_N and V_P. The resulting parameter variations are given by

$$\sigma'(P) = \frac{\sigma(P)}{\sqrt{8}}$$

References

1. M. Alioto, G. Palumbo, Power-aware design techniques for nanometer MOS current-mode logic gates: a design framework. IEEE Circuits Syst. Mag. **6**, 42–61 (2006)
2. M. Alioto, G. Palumbo, S. Pennisi, Modelling of source-coupled logic gates. Int. J. Circuit Theory Appl. **30**, 459–477 (2002)
3. T.C. Banwell, A. Jayakumar, Exact analytical solution for current flow through diode with series resistance. IEE Electron. Lett. **36**, 291–292 (2000)
4. D.A. Barry, P.J. Culligan-Hensley, S.J. Barry, Real values of the w-function. ACM Trans. Math. Softw. **21**(2), 161–171 (1995)
5. S. Bruma, Impact of on-chip process variations on MCML performance, in *IEEE International Conference on Systems-on-Chip (ICSOC)*, September 2003
6. M. Bucher, C. Enz, F. Krummenacher, J.M. Sallese, C. Lallement, A.S. Porret, The EKV 3.0 compact MOS transistor model: accounting for deep-submicron aspects, in *MSM*, vol. 2002 (2002), pp. 670–673
7. R.M. Corless, G.H. Gonnet, D.E.G. Hare, D.J. Jeffrey, D.E. Knuth, On the lambert w function. Adv. Comput. Math. **5**, 329–359 (1993)
8. C.C. Enz, F. Krummenacher, E.A. Vittoz, An analytical MOS transistor model valid in all regions of operation and dedicated to low-voltage and low-current applications. Analog Integr. Circuits Signal Process. **8**(1), 83–114 (1995)
9. D. Foty, *MOSFET Modeling with SPICE* (Prentice Hall, Upper Saddle River, 1997)
10. J.R. Hauser, Noise margin criteria for digital logic circuits. IEEE Trans. Educ. **36**, 363–368 (1993)
11. J. Lohstroh, E. Seevinck, J. De Groot, Worst-case static noise margin criteria for logic circuits and their mathematical equivalence. IEEE J. Solid-State Circuits **18**, 803–807 (1983)
12. S. Nadarajah, S. Kotz, Exact distribution of the max/min of two gaussian random variables. IEEE Trans. Very Large Scale Integr. Syst. **16**(2), 210–212 (2008)
13. M.J.M. Pelgrom, A.C.J. Duinmaijer, A.P.G. Welbers, Matching properties of MOS transistors. IEEE J. Solid-State Circuits **24**(5), 1433–1439 (1989)
14. T. Sakurai, A. Richard Newton, Alpha-power law MOSFET model and its applications to CMOS inverter delay and other formulas. IEEE J. Solid-State Circuits **25**, 584–594 (1990)
15. T. Sakurai, A. Richard Newton, A simple MOSFET model for circuit analysis. IEEE Trans. Electron. Devices **38**, 887–894 (1991)
16. U. Schaper, C.G. Linnenbank, R. Thewes, Precise characterization of long-distance mismatch of CMOS devices. IEEE Trans. Semicond. Manuf. **14**(4), 311–317 (2001)
17. A. Tajalli, E. Vittoz, Y. Leblebici, E.J. Brauer, Ultra-low power subthreshold current-mode logic utilising PMOS load device. IEE Electron. Lett. **43**(17), 911–913 (2007)
18. A. Tajalli, Y. Leblebici, E.J. Brauer, Implementing ultra-high-value floating tunable CMOS resistors. IEE Electron. Lett. **44**(5), 349–351 (2008)

References

Chapter 3
Design of MOS Current-Mode Logic Cells

3.1 Design Methodology for MCML Logic Gates

The design of MCML gates involves primarily the choice of a number of parameters, such as voltage swing and bias current, constrained by minimum noise margin specifications, resulting in values for transistor sizes. The quality of a design point can be quantified by a number of different metrics, most importantly delay and power dissipation.

The first step in solving the problem of optimally designing MCML gates consists in establishing the relationship between the design parameters and the quantities that we are trying to optimize.

3.1.1 *Trade-Offs*

As analyzed in Sect. 2.3.5, the delay of an MCML gate is essentially proportional to the time constant τ, which is the product of the equivalent load resistance by the total capacitance at the output nodes.

$$\frac{t_d}{\ln 2} = \tau = R_L \cdot C_{tot} = \frac{V_{sw}}{I_{SS}} C_{tot} \tag{3.1}$$

The total capacitance at the output is the sum of the parasitics due to the differential pair transistors, the parasitics due to the pull-up device, and any external load capacitance.

$$C_{tot} = 2C_{gd} + C_{db} + C_P + C_L \tag{3.2}$$

S. Badel et al., *Design Automation for Differential MOS Current-Mode Logic Circuits*, https://doi.org/10.1007/978-3-319-91307-0_3

The gate-to-drain and drain-to-bulk capacitances of an MOS transistor may be, to a first approximation, considered as proportional to the transistor's width

$$2C_{gd} + C_{db} \propto W_N \tag{3.3}$$

Note that, in multi-level gates, the parasitics associated with internal nodes are due for the most part to the junction capacitances of the NMOS transistors in the logic network. Therefore, (3.3) still holds for multi-level gates, with a different proportionality factor.

The parasitics associated with the load device (if implemented with a PMOS transistor biased in the linear region) are mainly due to gate-to-drain and drain-to-bulk capacitances, as in the case of the NMOS devices. From (2.64), we calculate β_P as

$$\begin{aligned}\beta_P &= \frac{I_{SS}}{V_{sw} \cdot \left(V_{DD} - V_P - V_{TP} - \frac{n}{2} V_{sw}\right)} \\ &\approx \frac{I_{SS}}{V_{sw} \cdot (V_{DD} - V_P - V_{TP})}\end{aligned} \tag{3.4}$$

The parasitics contributed by the load devices may also be approximated as proportional to their width, therefore

$$C_P \propto \frac{1}{R_L} = \frac{I_{SS}}{V_{sw}} \tag{3.5}$$

A remark can be made at this point, concerning the implementation of load devices. Though we are assuming in this discussion that load devices are implemented as PMOS transistors, for large values of I_{SS} the use of passive resistors may be more effective as discussed in Sect. 2.3.1. In that event, the parasitics associated with the load devices would have a different relationship to the effective resistance. In a passive resistor, the total parasitic capacitance is proportional to the device area, while the effective resistance is proportional to the L/W ratio. Therefore, the C_P component is proportional to V_{sw}/I_{SS} in this case, resulting in different tradeoffs.

Regarding transistors in the differential pairs, their sizes are dictated by noise margin requirements. Due to increasing subthreshold conduction reducing the switching efficiency of the differential pairs, the worst case noise margin occurs at the highest operating temperature. Therefore, by reversing (2.33), this corresponds to a ξ value of

$$\xi = 1 + \frac{NM}{4nU_{T(\max)}} \cdot \left(1 + \sqrt{1 + \frac{8nU_{T(\max)}}{NM}}\right) \tag{3.6}$$

where $U_{T(\max)}$ is the thermal voltage at the highest operating temperature. Through (2.32), we obtain

$$W_N \propto I_S = \frac{I_{SS}}{2\left(\frac{V_{sw}}{2nU_{T(\max)}} - \xi\right)^2} \tag{3.7}$$

Replacing back these results, we can calculate the gate delay as a function of the tail current, voltage swing, and load capacitance as

$$\frac{t_d}{\ln 2} = A_N \cdot \frac{V_{sw}}{\left(\frac{V_{sw}}{2nU_{T(\max)}} - \xi\right)^2} + A_P + \frac{C_L}{I_{SS}} V_{sw} \tag{3.8}$$

where A_N and A_P are technology-dependent proportionality constants, relating the parasitics to the transistor sizes. The propagation delay is therefore the sum of several components:

- a constant component due to load devices,
- a component due to external loading, that increases proportionally to V_{sw}, and
- a component due to the differential pair devices, that decreases with V_{sw}.

According to this formula, the intrinsic delay of an MCML gate (i.e., when considering an unloaded gate, that is, $C_L = 0$) is independent of the tail current. This is because, both the differential pair and the pull-up devices have to be sized proportionally to I_{SS} in order to keep the gain (and thus the noise margin, at constant voltage swing) constant; therefore, all parasitics scale proportionally to I_{SS}, while the resistance R_L scales in inverse proportion and the time constant does not change. The voltage swing, however, has a strong impact on the delay: as the voltage swing is decreased, the gain needs to be increased to keep the noise margin constant. The increase in transistor size is faster than the decrease in voltage swing, thereby increasing the time constant.

Formula (3.8) is compared against simulation results in Fig. 3.1. In this figure, the simulated delay of MCML buffers sized for different noise margins (50 and 100 mV), voltage swings (ranging from 140 to 500 mV), and load capacitances appears with symbols, and the different curves represent the delay given by expression (3.8) for the same parameters. The parameters A_N and A_P are obtained by fitting, but are the same for all the curves displayed on this graph. The graphs show a good agreement between the theoretical and simulation results.

An observation can be made regarding the relationship between noise margin and voltage swing. It is clear from (2.32) that the theoretical maximum value for ξ is $V_{sw}/(2nU_T)$. Therefore, the maximum noise margin is

$$NM_{\max} = V_{sw} \cdot \sqrt{1 - \frac{2nU_T}{V_{sw}}} - 2nU_T \tanh^{-1}\sqrt{1 - \frac{2nU_T}{V_{sw}}} \tag{3.9}$$

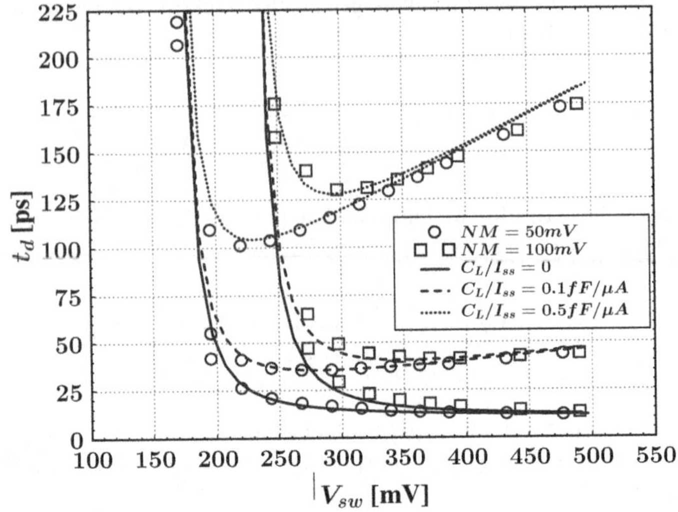

Fig. 3.1 Simulated propagation delay of an MCML buffer as a function of the noise margin, voltage swing, and load capacitance. Symbols denote simulation results obtained with 90 nm CMOS technology, with $I_{SS} = 50\,\mu\text{A}$, V_{DD}=1.2 V. Values of A_N and A_P obtained by fitting are 88 ps mV and 15.3 ps, respectively

which is, obviously, always smaller than V_{sw}. This explains why the voltage swing appears in the denominator in (3.8): for a given noise margin specification, there is an absolute minimum voltage swing and the transconductance (hence, transistor sizes) must increase towards infinity when the voltage swing approaches this minimum value. Interestingly also, $NM_{\max}$ is null when $V_{sw} = 2nU_T$, which implies that the voltage swing must always be larger than the value $2nU_T$ to have a positive noise margin.

When the gate is loaded ($C_L \neq 0$), the portion of the delay due to the external load is proportional to V_{sw}/I_{SS}. Therefore, the overall delay is the sum of an intrinsic component which decreases with V_{sw}, and an extrinsic component which increases with V_{sw}. There is an optimum value which depends on C_L/I_{SS}. When C_L/I_{SS} is large, the relative importance of the external load is large, and the optimum V_{sw} will be lower. At the limit, when $C_L/I_{SS} \to \infty$, the optimum voltage swing will be the minimum value given by (3.9). Conversely, when $C_L/I_{SS} \to 0$, the optimum voltage swing will be the largest possible value.

3.1.2 *Practical Limits of the Voltage Swing*

As it has been discussed, for a heavily loaded gate the optimum voltage swing tends to decrease. For a given noise margin specification, there is an absolute lower limit for the voltage swing according to Eq. (3.9). Using approximation (2.33), the minimum voltage swing can be expressed as

$$V_{sw(\min)} \approx 2nU_{T(\max)} + \frac{NM}{2}\left(1+\sqrt{1+\frac{8nU_{T(\max)}}{NM}}\right) \approx NM + 4nU_{T(\max)} \tag{3.10}$$

This is a limit value which allows to achieve the given noise margin when the transistor size is increased to infinity. The optimum voltage swing will always be larger than this value, for any realistic loading condition. Practically, the available area will also limit the voltage swing to a value higher than $V_{sw(\min)}$.

On the other hand, when considering a lightly loaded gate, the optimum voltage swing tends to increase. The maximum voltage swing is limited by two main factors.

First, the relationship (3.7) holds when the differential pair is operating in saturation. In order for the pair to be saturated at all times, the voltage swing must be smaller than the threshold voltage of the NMOS transistors

$$V_{sw(\max)} = V_{TN,\min} \tag{3.11}$$

Beyond that value, the gate delay, that was decreasing with increasing voltage swing, tends to saturate. In practice, it is still useful to increase the voltage swing somewhat further, until the noise margin starts to degrade. From this point of view, it might be beneficial to use NMOS transistors with large threshold voltage when further increase of the maximum voltage swing is desired.

The second limiter for the voltage swing is the pull-up device characteristic. When the pull-up devices are implemented as PMOS transistor biased in the linear region, as we are assuming in this discussion, the voltage swing has to be low enough to avoid the device entering saturation. Practically, this means that the maximum voltage swing must satisfy

$$V_{sw} < V_{ds,sat} = \frac{(V_{DD} - V_P) - V_{TP}}{n} \tag{3.12}$$

Practically, transistor sizes must be chosen so that $V_P \geq 0$ in the worst-case case conditions, i.e.

$$\left(\frac{W}{L}\right)_P = \frac{I_{SS}}{k'_{P(\min)} V_{sw}\left[V_{DD(\min)} - V_{TP(\max)} - \frac{n}{2} V_{sw}\right]} \tag{3.13}$$

where $k'_P = \mu_P C_{ox}$. However, the smallest drain-source saturation voltage occurs when V_P is the largest, that is, when V_{TP} is minimum and k'_P maximum

$$V_{P(\max)} = V_{DD} - V_{TP(\min)} - \frac{n}{2} V_{sw} - \frac{I_{SS}}{k'_{P(\max)} \left(\frac{W}{L}\right)_P V_{sw}} \tag{3.14}$$

Combining (3.13) and (3.14), we get the maximum value of V_{sw} as

$$V_{sw(\max)} = V_{ds,sat(\min)} = \frac{V_{DD(\min)} - V_{TP(\max)}}{n} \cdot 2 \frac{k'_{P(\min)}/k'_{P(\max)}}{1 + \left[k'_{P(\min)}/k'_{P(\max)}\right]} \tag{3.15}$$

In practice however, biasing the load devices at the limit of saturation makes the voltage swing very sensitive to variations in the effective tail current, as observed in Sect. 2.6. For this reason, it is desirable to limit the voltage swing to value even lower than given by (3.15).

From these considerations, it follows that the use of a lower threshold voltage for the PMOS load devices is beneficial, allowing to increase the maximum voltage swing and/or to reduce their dimensions so as to decrease the gate delays. The use of a higher supply voltage provides the same benefits; however at the expense of power dissipation.

3.2 MCML Latches and Flip-Flops

3.2.1 MCML Memory Element

The basic static memory element is a positive feedback loop. Because MCML logic can be non-inverting as well as inverting, by the choice of differential signal polarity, a memory element can be realized with a single buffer gate with the output fed back to the input in a non-inverting fashion, as depicted in Fig. 3.2a.

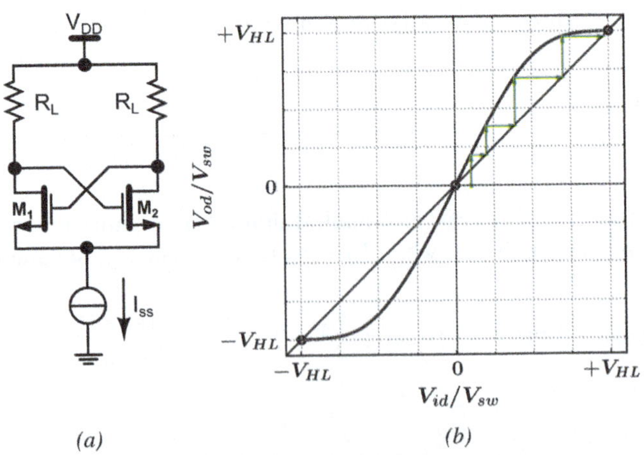

Fig. 3.2 Basic MCML static memory element (**a**) circuit schematic (**b**) operation

The operation of this circuit can be understood by noticing that the output of the buffer is also the input to itself, therefore stable operation can only be achieved when $V_{in} = V_{out}$, which defines three points on the transfer curve: $V_{in} = V_{out} = \pm V_{HL}$, and $V_{in} = V_{out} = 0$, as shown in Fig. 3.2b. In the normal operating state, the origin $V_{in} = V_{out} = 0$ is an *unstable* operating point. This is because the gain at this point is larger than one (a necessary condition to have positive noise margin), therefore, should the gate settle at this operating point, any perturbation would be amplified by the positive feedback and drive the circuit to either one of the two stable operating points. As shown in Fig. 3.2b with arrows, the path corresponding to successive approximations of the transfer function, the circuit is led to settle to $+V_{HL}$ for any positive initial differential voltage; conversely, any negative initial condition will drive the circuit to the $-V_{HL}$ operating point.

The memory effect is thus obtained by the presence of two stable operating points. The storing of a value in the element is achieved by one of two means:

- cut the feedback, change the value of the storage nodes, and then re-enable the feedback
- perturb the circuit to suppress one stable operating point, forcing the circuit to settle at the other operating point

3.2.2 *MCML Latch*

The MCML latch is constructed based on the first mode of writing into the memory element: a memory element will be selectively turned on when retention of the logic state is desired. As depicted in Fig. 3.3a, the circuit is equivalent to a buffer (driven

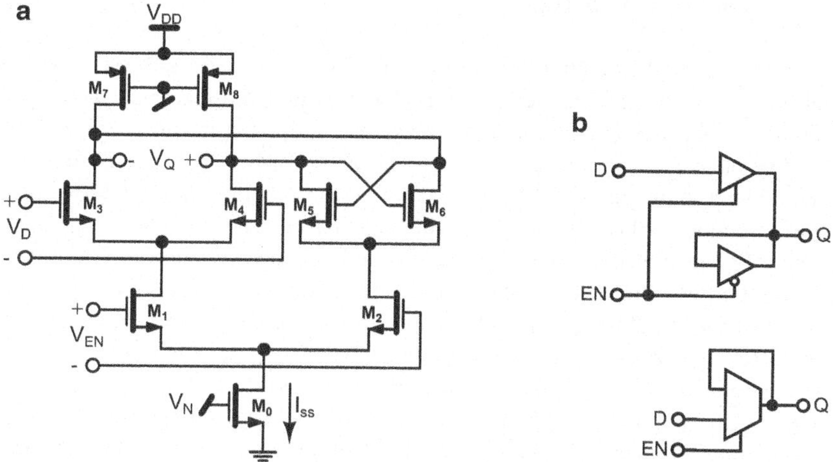

Fig. 3.3 (**a**) Circuit schematic of an MCML multiplexer-based latch. (**b**) Equivalent circuit representations

by V_D) and a memory element connected in parallel, with only one of the two active at a given time due to the presence of a single tail current switched by the V_{EN} signal into one of the two pairs. Therefore, when EN is positive, the buffer will be active and the output will be equal to D: the circuit is transparent. On the other hand, when EN switches to a negative value, the feedback memory element is activated and the output will settle and rest at one of the two stable states.

This circuit is formally a latch, because it is *level-sensitive*—it is either transparent to the input or in latch state, depending on the level of the EN signal.

3.2.2.1 Dynamic Operation

The dynamics of the latch circuit depend on the particular input transitions. Usually, one is interested in the delay from input to output—this delay in a latch can arise from two different transitions:

- V_D is changing while the latch is in transparent state: the delay $t_{D\rightarrow Q}$ is from D to Q.
- V_D changes while the latch is opaque, but V_Q changes only later when the latch is switching to transparent state: the delay $t_{EN\rightarrow Q}$ is from EN to Q.

In both cases, assuming that the voltage at the Q node is sufficiently stable when the switching occurs, the delay will by all means be equal to the corresponding delay of the multi-stage logic network. Therefore, the EN-to-Q delay will be slightly higher than the D-to-Q delay, and both will be slightly higher than in a MUX2 gate because of the additional loading at the output due to the gate capacitances of the cross-coupled transistors M_5 and M_6.

3.2.2.2 Setup and Hold Times

The switching of the latch from transparent into latch state deserves particular attention. When the memory element is activated and the feedback restored, the final output state will depend on the initial input state—that is, whether the Q node is positive or negative at the instant of the transition. Therefore, the D input must switch no later than one $t_{D\rightarrow Q}$ delay time before the EN input, in order to allow the Q node to change polarity before being latched. This defines the *setup time* t_{setup} of the latch. Because of this effect, the data to be latched can effectively be removed *earlier* than the EN transition, since the Q node is retaining the correct value for an amount of time equal to $t_{D\rightarrow Q}$. This defines the *hold time* t_{hold} of the latch, which is therefore negative, and we obviously have $t_{hold} = -t_{setup}$ (Fig. 3.4).

These definitions of setup and hold time are based on functional operation, that is, the conditions under which the circuit will or will not be able to produce the expected output. In practice, setup and hold times are defined based on timing considerations. This is because, when inputs are switched at short interval, even though the circuit might eventually produce correct output, the delays can increase

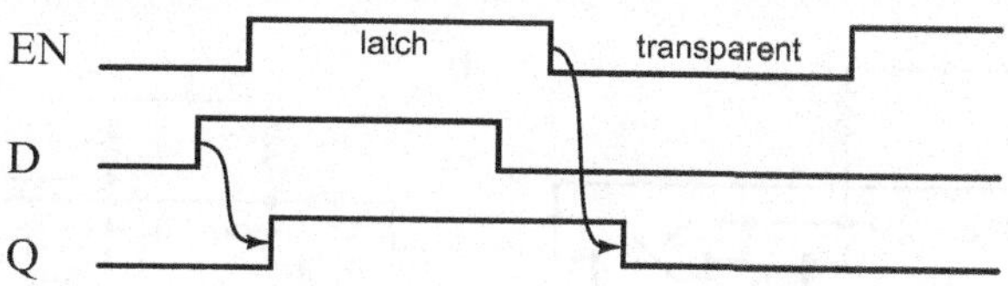

Fig. 3.4 Timing diagram for a latch. The switching of the D inputs causes an output transition when the latch is transparent; the switching of the EN input causes an output transition when the data has changed during the latch period

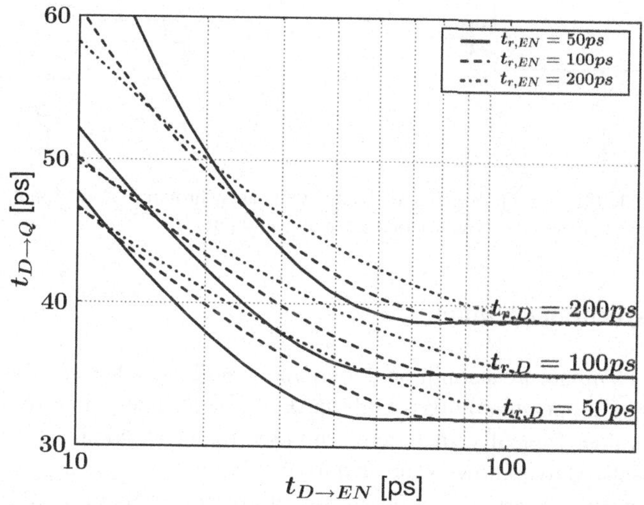

Fig. 3.5 Simulated $t_{D\to Q}$ delay for an MCML latch. The increase in delay for low $t_{D\to EN}$ defines the setup time of the latch for this transition. Simulations carried on 180 nm CMOS technology with parameters: $V_{sw} = 400\,\text{mV}$, $I_{SS} = 50\,\mu\text{A}$, NM = 120 mV

considerably (see Fig. 3.5). Therefore, the common definition for setup and hold times (as well as other timing metrics for sequential circuits) is: minimum time interval between two events that does not cause an increase in delay larger than a defined value—usually, a percentage of the nominal delay, measured when the time interval is very large.

Figure 3.5 displays simulated delay $t_{D\to Q}$ of an MCML latch, with respect to the relative time $t_{D\to EN}$ between the switching of the D and EN signals. As expected, for large values of $t_{D\to EN}$, the delays are constant, while they increase for smaller values. In this case, the feedback element is being activated in a state too close to the unstable state, causing a *metastable* behavior which can last for an extremely long period of time until it finally settles to a stable operating point.

Note that, in the latch circuit as shown in Fig. 3.3a, the setup and hold times are dependent on the settling of the Q node, and therefore depend on the loading of this node. For this reason, latches should have their output buffered, in order to decouple the Q node from external parasitics.

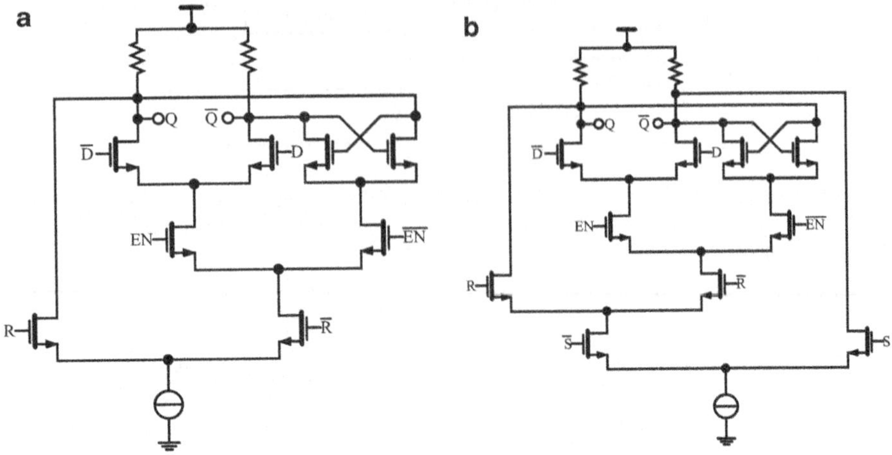

Fig. 3.6 (**a**) MCML multiplexer-based latch with asynchronous reset input. (**b**) MCML multiplexer-based latch with asynchronous set and reset inputs

3.2.2.3 Asynchronous Inputs

Asynchronous inputs are a common feature integrated in latches and flip-flops. They are transparent inputs, that bypass both the EN and D states and directly affect the memorized value. Typically, such inputs are used for clear and preset, when a circuit must be initialized to a known state at startup or reset.

Implementation of an asynchronous set and reset signals in an MCML latch is illustrated in Fig. 3.6a, b. The circuit in Fig. 3.6a implements asynchronous reset, by diverting the tail current to the positive output node—effectively pulling the output low—when reset is active. Note that, because the R differential pair is located below the EN in the logic network, it has priority over it and thus the R input is asynchronous. By doing the opposite, that is, effectively replacing the buffer function at the D input by a logic function of multiple inputs, one obtains additional synchronous inputs. In Fig. 3.6b, another input is added to implement asynchronous preset. In this configuration, if both S and R are asserted simultaneously, S takes precedence because it is located at the lowest level of the logic network. Note that the clear or preset behavior of an input can be inverted if the data input and outputs are also inverted. By this transformation, the same circuit can realize asynchronous set or reset depending on the configuration.

3.2.3 Master–Slave MCML Latch

A master–slave latch (MSL) is obtained by cascading two latches with different polarities of the EN signal (Fig. 3.7). The purpose of combining to latches in a

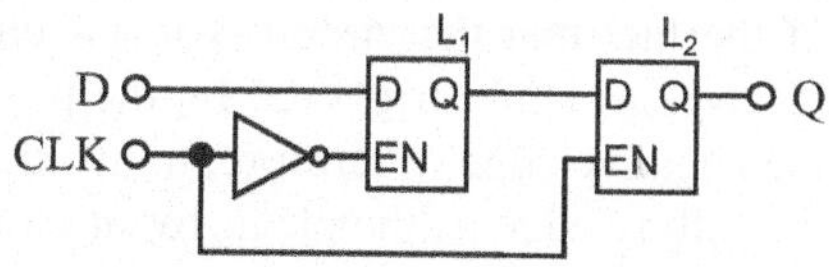

Fig. 3.7 Master–slave latch

master–slave fashion is to eliminate the transparency window existing in the latch operation, in order to obtain an edge-triggered memory element.

The MSL operates as follows:

- during the negative phase of the clock, the master latch L_1 is transparent while the slave latch L_2 is opaque. The value of D is transferred to the input of L_2.
- during the positive phase of the clock, the master latch L_1 captures the last value that D had in the previous phase, while the slave latch L_2 becomes transparent and transfers that value to the output.

Effectively, the value of D is transferred from the input to the output of the master–slave latch at the rising edge of the clock. It should be noted, though, that the operation is level-sensitive, based on the two phases of the clock signal. This property makes the master–slave latch a very tolerant circuit with respect to the rise and fall times of the clock signal; practically, as long as the data is stable, the MSL will capture the correct value even with an extremely slowly switching clock signal. However, due to the presence of two latch circuits, it is also a power- and area-hungry circuit.

3.2.3.1 Setup and Hold Times

The timing behavior of the master–slave latch is dictated by the operation of the latch circuit. At the positive edge of the clock, the master latch L_1 is switching from transparent into latch mode, and the data input must satisfy a corresponding setup time constraint. At the same moment, the latch L_2 is becoming transparent, and the data is transferred to the output with a delay $t_{CLK \to Q}$ relative to the clock edge. Since the input to the second latch does not change during the next clock period, there is no constraint at the falling edge of the clock. Note that the clock-to-Q delay is caused by a combination of $t_{D \to Q}$ of L_1 and $t_{EN \to Q}$ of L_2; however as it was noted previously, the transition of the second latch is hardly sensitive to the setup time, which is dominated by the first latch. The relative size of both latches will therefore determine a tradeoff between setup and delay times: a larger L_2 device will have more drive capability, but present more load to the previous stage and hence increase its setup time.

Note that, when the latches are used in a master–slave configuration, the first latch has its output decoupled from external parasitics by the second latch; therefore, its setup time is not dependent on the loading, and no buffering is needed. The second latch, however, stores the data at node Q at the falling edge of the clock, and setup constraint applies between Q and the falling edge of CLK. This is a consequence of the fact that the circuit is level-sensitive : though from the outside, the circuit behaves as being triggered by the positive edge of the clock, it is also reacting to the negative edge. Therefore, to avoid the metastability or latching of wrong data at the falling edge of the clock for any loading conditions, the MSL output should be buffered. Note that, since the data at the input of the second latch is always stable long before the falling edge of the clock (since it is being held stable by the first latch), the stability is guaranteed for any clock-to-Q delay smaller than about one half of the clock period. In practice, a situation where the delay is larger than half of the clock period is extremely unlikely to arise, and it is acceptable not to buffer the output if proper care is taken.

3.2.3.2 Asynchronous Inputs

In a master slave latch, asynchronous inputs must affect the output value until the next rising edge of the clock. Because of this, both master and slave latches must react to the asynchronous inputs. To understand this, consider the situation where a pulse is applied on an asynchronous input, during either the positive or the negative phase of the clock:

- during the negative phase of the clock, the slave latch L_2 is opaque. Therefore, for the output to be affected, the asynchronous input needs to be applied to the slave latch.
- during the positive phase of the clock, L_2 is transparent. If the asynchronous input affects L_2 only, the output is affected only as long as the asynchronous input is active; as soon as it is released, the output switches back to the previous value stored in the master latch L_1. Therefore, the asynchronous input must also be applied to the master latch.

3.2.4 MCML Flip-Flop

A flip-flop is an edge-sensitive circuit. It is different from a master–slave latch in that it is composed of a single memory element, and a pulse-generating unit that reacts to the transition of the clock and captures the input data during a very short window of time. As opposed to the master–slave latch which is level-sensitive, the flip-flop is really sensitive to the edge of the clock signal. The presence of only a single latch circuit may allow a flip-flop to have a simpler implementation than a MSL, at the expense of an increased sensitivity to the clock rise and fall times (Fig. 3.8).

Fig. 3.8 Logic representation of a flip-flop

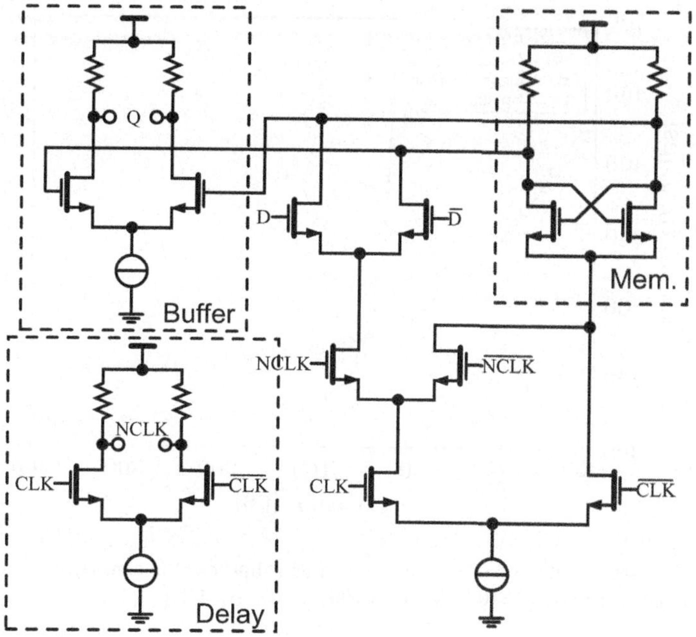

Fig. 3.9 Pulse generator-based MCML flip-flop

An implementation of a flip-flop circuit in MCML is depicted in Fig. 3.9. It is composed of a memory element, a pulse generator, and a buffer. The pulse generation is done by a logical AND between the clock signal and a delayed version of itself. This provides a short pulse of current to the differential pair driven by the *D* input, at the instant of clock transition. This pair being connected to the memory element, it will force the latched value to the value of *D*. The buffer isolates the memory element from external parasitics, in order to have a well-defined load and guarantee that the latch will be able to switch.

In this type of circuits, the clock-to-*Q* delay is typically larger than for a master–slave latch, because the path from clock to output goes through two stages: the core flip-flop and the buffer. However, since the transparency window is open for a short time *after* the clock transition, such a flip-flop usually has a smaller setup time, which compensates for the increased delay. This is illustrated with simulation results in Fig. 3.10, where the clock-to-output and data-to-output delays of both master–slave latch and flip-flop are displayed as a function of the data-to-clock delay (setup time). As it can be seen, the delay is lower for the master–slave latch, however it increases dramatically for setup times lower than about 100 ps. The flip-flop can tolerate much lower setup times, with a relatively smaller increase in delay. The resulting data-to-output delay, $t_{D\rightarrow Q} = t_{D\rightarrow CLK} + t_{CLK\rightarrow Q}$, can be lower with a flip-flop circuit, allowing a shorter clock period in a pipeline stage.

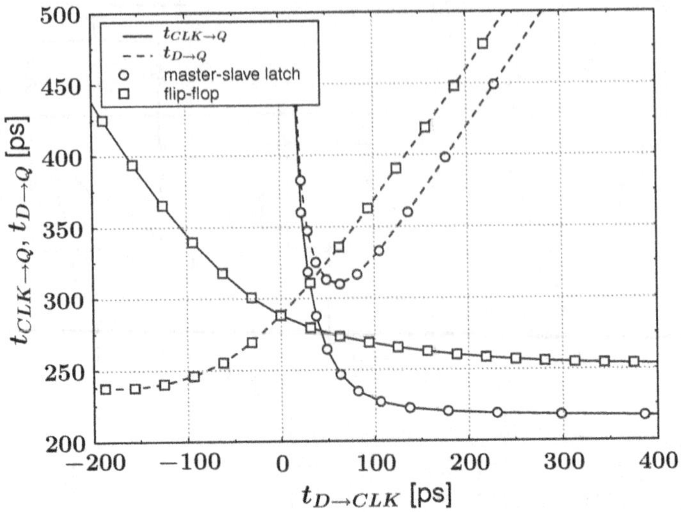

Fig. 3.10 Clock-to-Q delay of master–slave latch and flip-flop. Simulations carried in 180 nm CMOS technology, with $V_{sw} = 400\,\text{mV}$, $I_{SS} = 100\,\mu\text{A}$, $C_{load} = 50\,\text{fF}$

Fig. 3.11 Simplified MCML flip-flop circuit

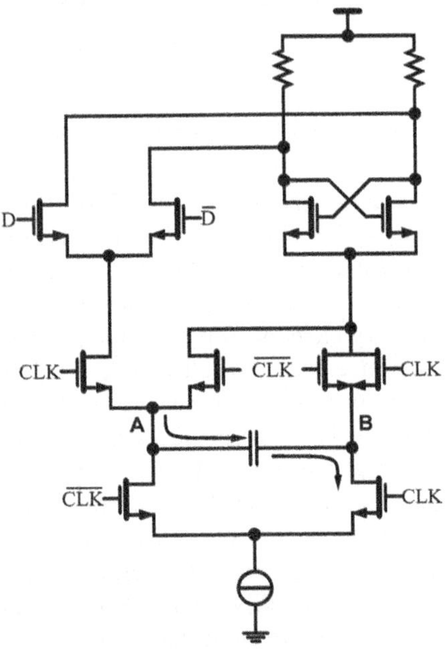

In order to save the power dissipation and area cost of the delay elements, the basic flip-flop circuit can be modified as shown in Fig. 3.11. In this variant, the current pulse needed to toggle the stored value is generated by charging/discharging of internal parasitics. The circuit operates as follows: when the clock signal toggles from negative to positive, the voltage at node *A* rises by one overdrive voltage. Since the clock is high, the charging current goes through the *D* differential pair, enabling the transparency of the circuit. At the same time, node *B* has to be pulled low, and this discharging current is subtracted from the latch current, partially disabling the latching process, which eases the toggling of the output. The value of the capacitor can be tuned to adjust the duration of the transparency window.

3.2.4.1 Asynchronous Inputs

Asynchronous inputs can be added to an MCML flip-flop by the same modifications as in the latch circuit. Note that, compared to a master–slave latch, a flip-flop incurs less overhead for the addition of asynchronous inputs since only a single stage needs to be modified. However, the basic flip-flip circuit already needs three levels of stacked differential pairs in the core flip-flop; adding asynchronous inputs requires increasing the number of levels, and as it has been discussed with multi-stage logic networks, this implies an increase of transistor sizes and thus speed and area penalties.

Alternative realizations of the MCML flip-flop allow to move the memory element out of the pulse-generating tree, suppressing the need to increase transistor sizes. A possible implementation is shown in Fig. 3.12. In this circuit, the memory element is separated from the pulse-generation circuitry. This circuit is as fast as a regular flip-flop, and the area penalty is minimal; however, the need for an additional current source results in increased power consumption.

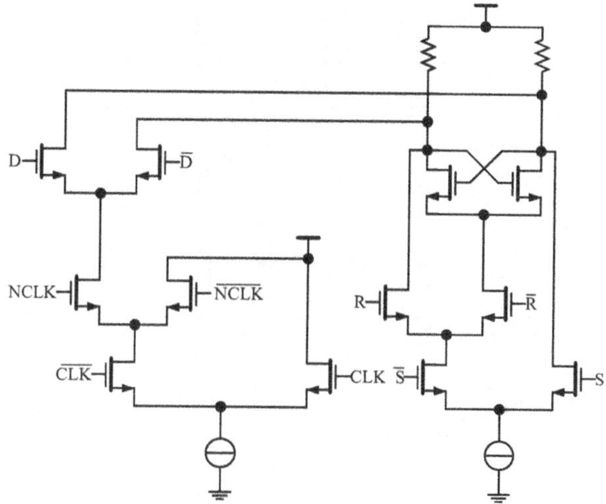

Fig. 3.12 MCML flip-flop with asynchronous set/reset inputs

3.2.5 Dual Edge-Triggered Elements

Dual edge-triggering is one technique allowing to decrease the clock frequency in a circuit, relaxing the constraints on the clock distribution network. Dual edge-triggered elements store data at both rising and falling edges of the clock signal, therefore allowing the clock frequency to be scaled by a factor of 2 with the same data throughput, with the drawback that the operation becomes sensitive to the duty cycle of the clock signal.

Both latch and flip-flop circuits can be modified for dual-edge triggering. Implementations with MCML elements are pictured in Fig. 3.13a, b. For the latch implementation, two latches have their outputs time-multiplexed, so that each one is transparent during a different half of the clock period, while the multiplexer transfers at the output the value stored in the other latch, captured at the previous clock transition. For the flip-flop, due to the circuit's symmetrical behavior, it can with very little modification be made sensitive to both clock edges, by merely replacing the logical AND between the clock signal and the delayed version of itself, providing the capturing pulse, by a logical XOR.

Interestingly, flip-flop circuits provide very efficient implementations of dual-edge triggered elements. In both cases, the overheads are rather low for turning single-edge triggered circuits into dual-edge triggered counterparts: for the flip-flop, the overhead is of one or two transistors, depending on the original implementation; for the latch circuit, a single mutliplexer has to be added. However, the latch circuit

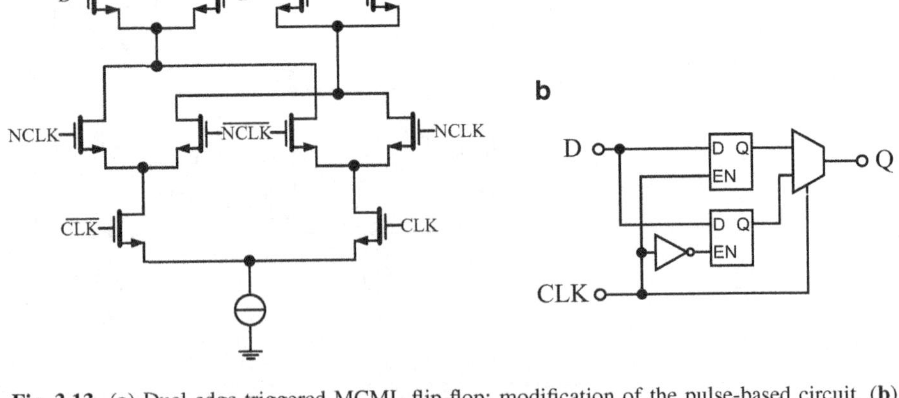

Fig. 3.13 (**a**) Dual edge-triggered MCML flip-flop: modification of the pulse-based circuit. (**b**) Dual edge-triggered MCML flip-flop: dual latch implementation

presents an increased input capacitance to the clock network, which is not the case for the flip-flop circuit. In terms of timing, the latch circuit incurs an additional delay in the clock-to-output path, through the multiplexer, which is also not the case in the flip-flop circuit.

3.3 Tri-State MCML Buffers

In digital logic circuits, it is often necessary to have several gates alternatively driving the same node. A common example is a bus allowing communication between several modules in a system. All modules may receive data from the bus at the same time, but only one may transmit data over the bus at a given time, thus requiring other modules's transmitters to disconnect from the bus. The most common way to achieve this is through the use of tri-state gates. Tri-state gates can drive the output either high or low—that is, they act as regular gates—when enabled, and leave their output floating when disabled; they have therefore three output states which are commonly referred to as 0 (or low), 1 (or high), and Z (high-impedance). Tri-state gates are easily designed in CMOS logic, because CMOS gates are implemented as networks of switches connecting the output to either the positive or the negative supply. Thus, it is easy to turn off the paths connecting both supplies to the output to drive it into high-impedance state. In contrast, MOS Current-Mode Logic gates have their outputs constantly connected to the positive supply via the load devices, which cannot be turned off. Moreover, the reduced signal swings do not allow to drive MOS switches efficiently.

An effective way to isolate the outputs is to include series MOS switches. However, MOS switches need a gate-to-source voltage swing that is many times larger than the threshold voltage in order to be turned on and off efficiently. Since MCML signal swing is only a few hundreds of millivolts, a circuit to increase the voltage swing is needed to drive the MOS switches. Such a circuit is presented in Fig. 3.14a. It is basically a differential amplifier with single-ended output. The DC characteristic (Fig. 3.14b) shows that the output swing ranges from the supply voltage to as low as 200 mV, which is sufficient to drive the PMOS switches. One drawback of using such a circuit is that, because it is unbalanced, it produces current spikes in the power supply during switching.

A complete tri-state buffer using series PMOS switches is depicted in Fig. 3.15. It acts as a regular MCML buffer when EN is high, and its outputs are driven into high-impedance state when EN is low. It features two current sources which can be of different values, multiples of a unit current I_{SS}. Note that this circuit has increased parasitics at the output nodes compared to the simple buffer, due to the PMOS switches. Also, the series resistance of the switches is limiting the current to charge or discharge external parasitics; however, this resistance is usually very low compared to the resistance of the load devices and has little effect on the time constants. In Fig. 3.15b, the Z-mode feedthrough is plotted against frequency when EN is low. The Z-mode feedthrough is the fraction of the input signal that is

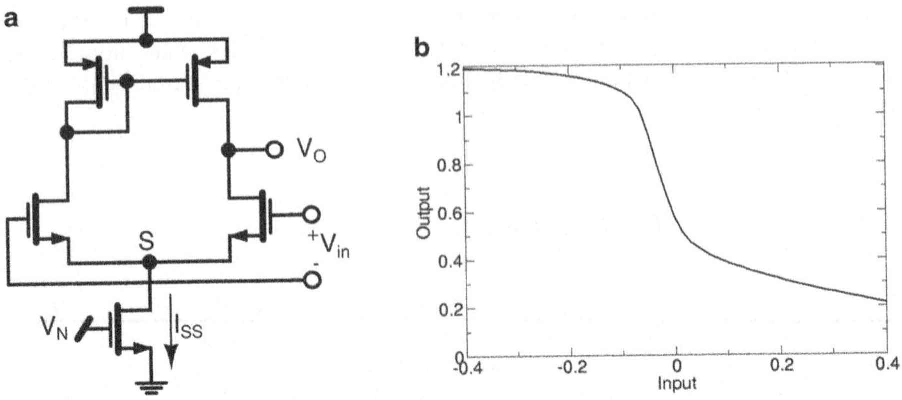

Fig. 3.14 (**a**) Driver circuit for MOS switches: the input differential voltage is turned into a single-ended voltage with increased voltage swing. (**b**) Sample simulated transfer characteristic of the switch driver circuit

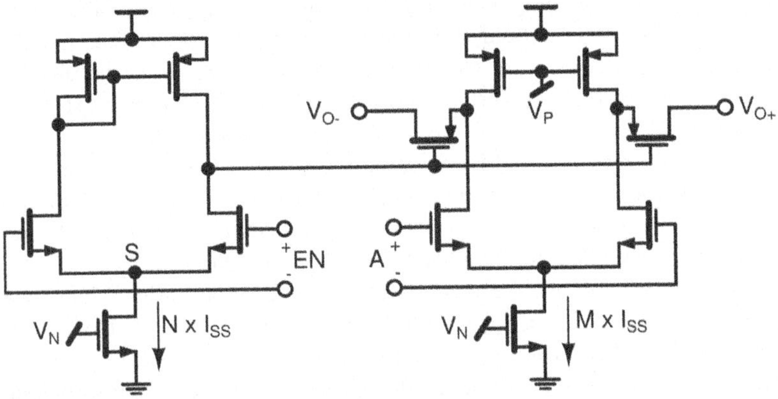

Fig. 3.15 Circuit schematic of the switch-based MCML tri-state buffer

transferred to the output when the gate is in high-impedance mode, and illustrates the ability of the circuit to isolate the output from the input. The peak feedthrough is lower than −50 dB, providing a good isolation (Fig. 3.16).

Another popular circuit that can provide high-impedance state is the switched voltage follower. A switched voltage follower can be conveniently adapted to MCML circuits by using a flipped voltage follower (Fig. 3.17a) as introduced in [5]. In contrast to a conventional source–follower, it makes use of a sinking current source (for a PMOS-type follower), which is readily available in MCML circuits. Also, the output of this circuit can be driven into high-impedance state when turning the current-source off, since the gate of the feedback transistor M_2 is driven high until it eventually turns off. In order to make the voltage levels compatible with MCML levels, it is needed to include a level shifter at the input, resulting in a

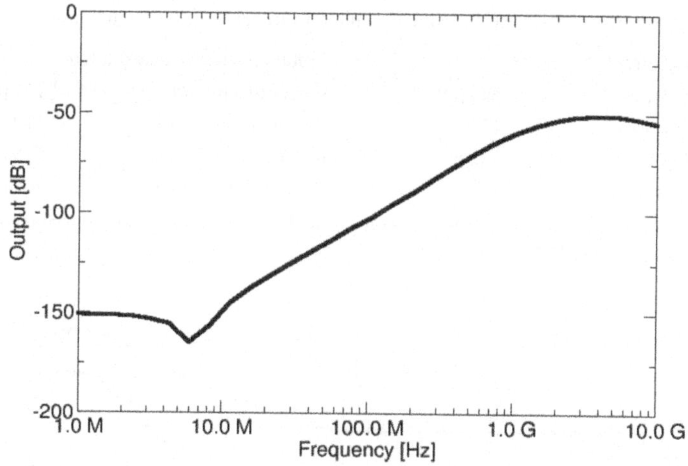

Fig. 3.16 Simulated Z-mode feedthrough of the switch-based MCML tri-state buffer

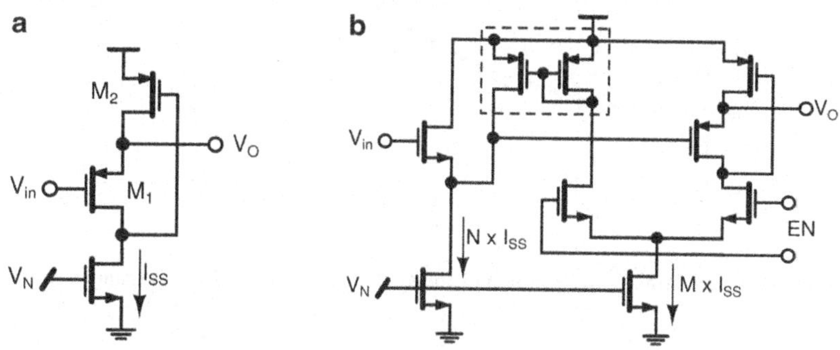

Fig. 3.17 (**a**) Circuit schematic of a flipped voltage follower. (**b**) Switched flipped voltage follower circuit. The tail current of the flipped voltage follower is switched by a differential pair. The input voltage is level-shifted by a conventional source follower

source follower-flipped voltage follower (SF-FVF) cascade. A differential pair is used to switch the bias current of the flipped voltage follower on and off, as shown in Fig. 3.17b. Because the gate of M_2 in the flipped voltage follower stage is floating, the circuit displays a poor isolation compared to the switched-based circuit. In order to improve the isolation, the circuit can be enhanced by mirroring the bias current diverted from the output stage by the differential pair when the circuit is in high-impedance state to pull the source of the input source follower stage high. With this modification, the switched voltage follower circuit achieves a level of isolation as good as the switched-based circuit. A complete tri-state buffer can be built by using two switched voltage follower circuits, for each polarity of the differential signal. Note that this circuit is also somewhat noisy because of the source followers.

Simulations show that, for similar characteristics, the voltage-follower circuit is more efficient than the switch-based circuit. The results summarized in Table 3.1

Table 3.1 Comparison of switch-based and voltage follower-based tri-state circuits

Switch-based circuit		Voltage follower-based circuit	
Power dissipation	564 μW	Power dissipation	191 μW
Power supply noise	122 μA	Power supply noise	119 μA
Input-output delay	332 ps	Input-output delay	315 ps
Enable-output delay	517 ps	Enable-output delay	491 ps

Simulations carried in 180 nm CMOS technology, with drivers sized to drive a 500 μm long bus line at 1 Gbps data rate

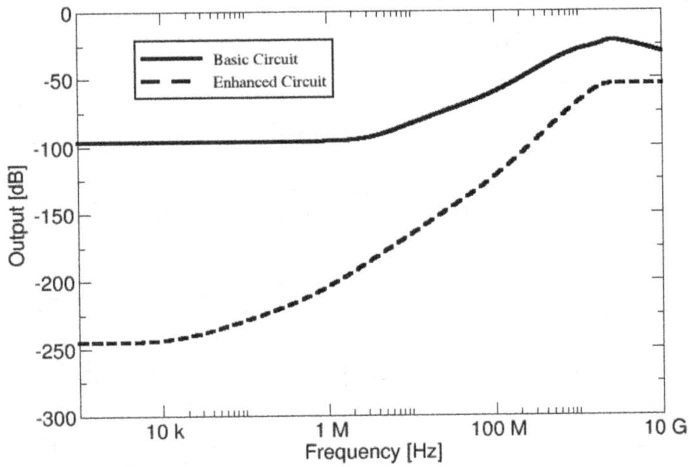

Fig. 3.18 Simulated Z-mode feedthrough of the switched voltage follower tri-state buffer

provide a comparison between the two types of circuits, both were designed for being able to drive a 500 μm long signal line at 1 Gbps data rate, in a 0.18 μm CMOS technology. As it can be seen, the voltage-follower circuit achieves the same performance dissipating less than one half of the power, while comparing the supply noise generated by both circuits, the performances are similar (Fig. 3.18).

3.4 High-Speed and Low-Power Techniques

3.4.1 Speed Enhancement with Peaking Techniques

Peaking is a technique used in high-frequency circuits to improve the time-domain response. The idea is to introduce a pole-zero couple in the transfer function, which can be achieved by various means. Passive elements usually offer the highest quality factor, hence the higher increase in performance, at the cost of high area occupation, which can be alleviated with active realizations.

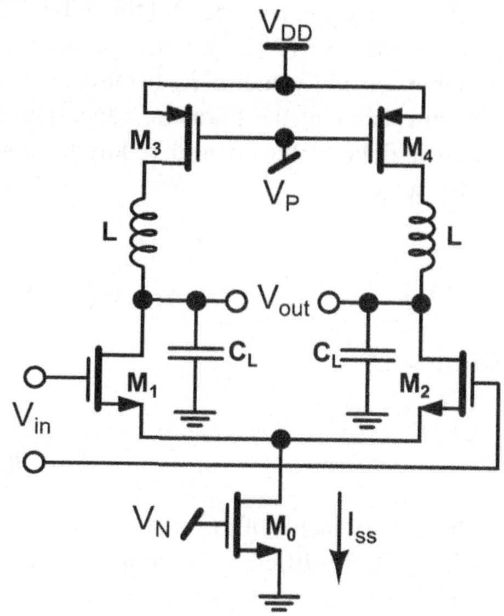

Fig. 3.19 Inductive peaking in an MCML gate

3.4.1.1 Passive Inductors

Peaking can be achieved through the insertion of an inductance at the output nodes [3, 4], as illustrated in Fig. 3.19. As it was done in the previous chapter to analyze the MCML buffer circuit, we can approximate the differential pair with a linear circuit and analyze the small-signal equivalent. Calculations yield the following transfer function

$$\frac{V_{out}}{V_{in}} = A_v \cdot \frac{1 + s \cdot \frac{L}{R}}{1 + s \cdot RC + s^2 \cdot LC} \tag{3.16}$$

The resulting system is of second order, with a zero at $s = -L/R = -(L \cdot I_{SS})/V_{sw}$. As for any second-order system, we can define a natural frequency ω_n and a damping ratio ξ

$$\omega_n = \frac{1}{\sqrt{L \cdot C}} \quad \text{and} \quad \xi = \frac{1}{2} R \sqrt{\frac{C}{L}} \tag{3.17}$$

The circuit is *critically damped* for $\xi = 1$, corresponding to $L = R^2C/4$. In that case, the gate delay can be approximated as $t_d = 0.573 \cdot RC$, which is already a 17% improvement over the first-order case. For values of L smaller than $R^2C/4$, the circuit will be *overdamped* and the delay will increase towards the limit value $t_d = 0.693 \cdot RC$ of the first-order system. For larger values of L, the circuit will become *underdamped* and the delay will drop, while the response will exhibit an increasing overshoot. The circuit will become *undamped* (i.e., present sustained oscillations) for the limit value $L = \infty$ and the delay will approach a minimum value of $0.5RC$, corresponding to a 27% decrease of the propagation delay. Note that this architecture is stable for any value of L. It must also be highlighted that this analysis does not take into account the parasitic capacitance contributed by the inductors, and thus practical delay values would be larger than those mentioned above.

By substituting $s' = s \cdot R \cdot C$ in the transfer function, it can be rewritten as

$$\frac{V_{out}}{A_V \cdot V_{in}} = \frac{1 + s' \cdot Q^2}{1 + s' + s'^2 \cdot Q^2} \tag{3.18}$$

where

$$Q = \frac{1}{R} \cdot \sqrt{\frac{L}{C}} \tag{3.19}$$

is the quality factor of the RLC resonator. The step response of the circuit is plotted in Fig. 3.20 for different Q values.

The main drawback of using passive inductances is that the inductance values needed to implement inductive peaking, in the range of tens to hundreds of nH, would require devices whose area is much larger than the gates themselves. This makes inductive peaking not practically implementable in standard-cell-based designs, which require a high integration density. Also, note that because the voltage across the inductor can be negative, the output nodes of the circuit can actually rise higher than V_{DD} when $Q > 1/2$ and there is overshoot.

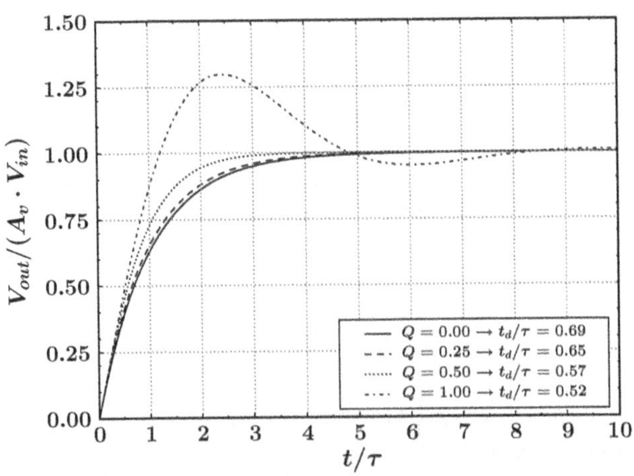

Fig. 3.20 Step response of the peaked MCML system for different Q values

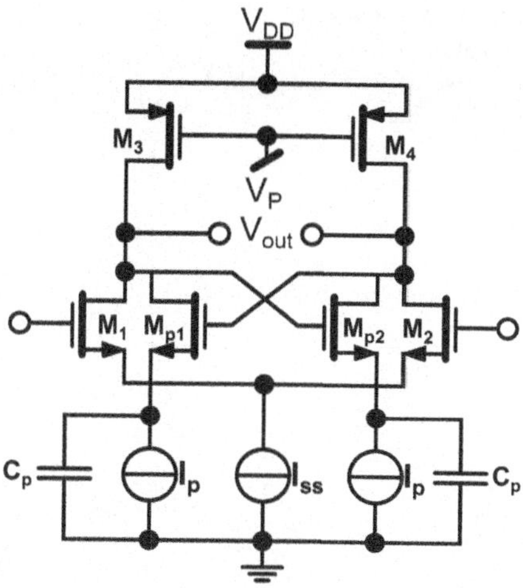

Fig. 3.21 Negative capacitance peaking in an MCML gate

3.4.1.2 Negative Capacitance

Instead of using inductors, peaking can be achieved by connecting negative capacitances in parallel with the load capacitance [1]. The negative capacitance can be realized by adding cross-coupled NMOS as illustrated in Fig. 3.21. In this configuration, the source of the cross-coupled NMOS is moving with the gate, which is moving opposite to the drain due to the cross-coupling. The result is that a positive and a negative peaks of current will be generated during switching, that will charge the source capacitances C_P; this differential current pulse will help the transition of the output nodes.

By performing the small-signal analysis of the circuit of Fig. 3.21, including the negative capacitance, the transfer function is found as

$$\frac{V_{out}}{V_{in}} = A_V \cdot \frac{1 + s \cdot \frac{C_P}{g_{mp}}}{1 + s \left[R_L \left(C_L - C_p \right) + \frac{C_P}{g_{mp}} \right] + s^2 \cdot R_L C_L \frac{C_P}{g_{mp}}} \tag{3.20}$$

which can be written as

$$\frac{V_{out}}{A_V \cdot V_{in}} = \frac{1 + s' \cdot \frac{\alpha}{\beta}}{1 + s' \cdot \left(1 - \alpha + \frac{\alpha}{\beta}\right) + s'^2 \cdot \frac{\alpha}{\beta}} \tag{3.21}$$

where

$$\alpha = \frac{C_P}{C_L}, \qquad \beta = \frac{R_L}{1/g_{mp}} = \frac{1}{A_v} \cdot \frac{g_{mp}}{g_m} \qquad \text{and} \qquad s' = \frac{s}{R_L C_L}$$

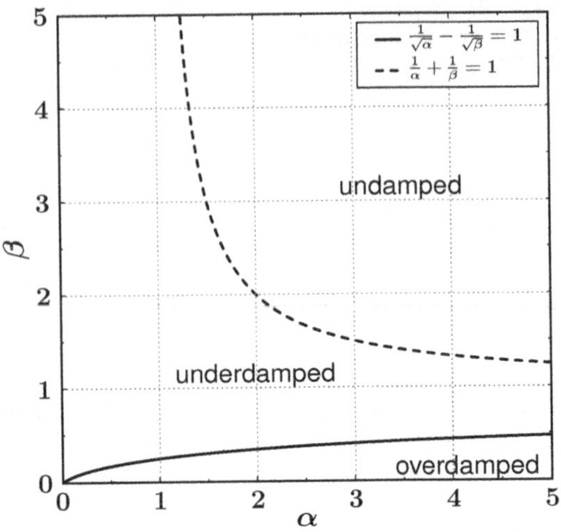

Fig. 3.22 Modes of operation in the negative-capacitance peaked MCML gate

As in the case of passive inductors, the circuit is of second order with one zero. However, in this case, there are two parameters, α and β, which control the natural frequency and damping ratio independently; these are given by

$$\omega_n' = \frac{\omega_n}{\sqrt{R_L C_L}} = \frac{\beta}{\alpha} \quad \text{and} \quad \xi' = \frac{\xi}{R_L C_L} = \frac{1}{2} \cdot \sqrt{\frac{\beta}{\alpha}} \cdot \left(1 - \alpha + \frac{\alpha}{\beta}\right) \tag{3.22}$$

The circuit will be critically damped for values of α and β satisfying $\xi = 1$, which results in

$$\frac{1}{\sqrt{\alpha}} - \frac{1}{\sqrt{\beta}} = 1 \tag{3.23}$$

For high values of α and β, the circuit can become undamped (oscillatory). The undamped region in the (α,β) plane is bounded by the curve

$$\frac{1}{\alpha} + \frac{1}{\beta} = 1 \tag{3.24}$$

This defines three regions in the (α,β) plane as illustrated in Fig. 3.22. The resulting delay and overshoot are dependent on the particular values of α and β as plotted in Figs. 3.23 and 3.24 as level curves. The region of interest, with $\alpha > 1$ and $\beta < 1$, offers a delay lower than the first-order system by 30–50%, with an overshoot lower than about 30% for a wide range of α values.

It should be noted that the cross-coupled devices contribute some parasitics at the output nodes, and therefore practical delay values are higher than the theoretical values displayed in Fig. 3.23. Also, it is important to outline that the negative capacitance circuit requires a bias current of $2 \cdot I_p$, which results in an increase of

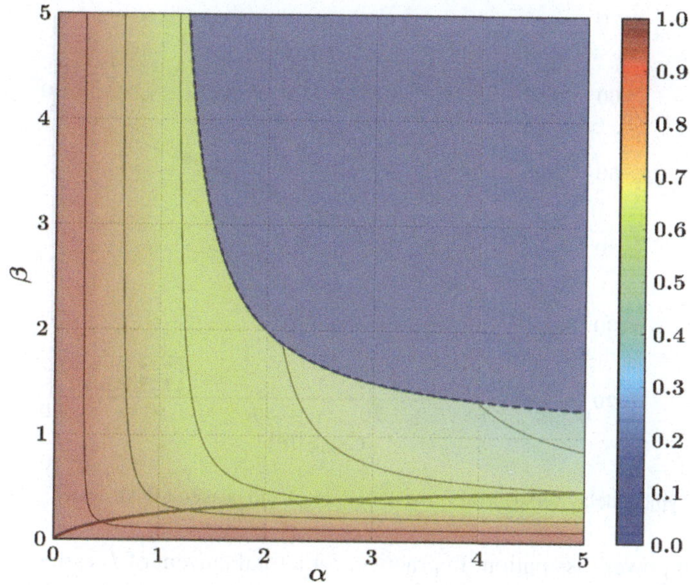

Fig. 3.23 Relative delay ($t_d/0.69RC$) in the negative-capacitance peaked MCML system as a function of α and β

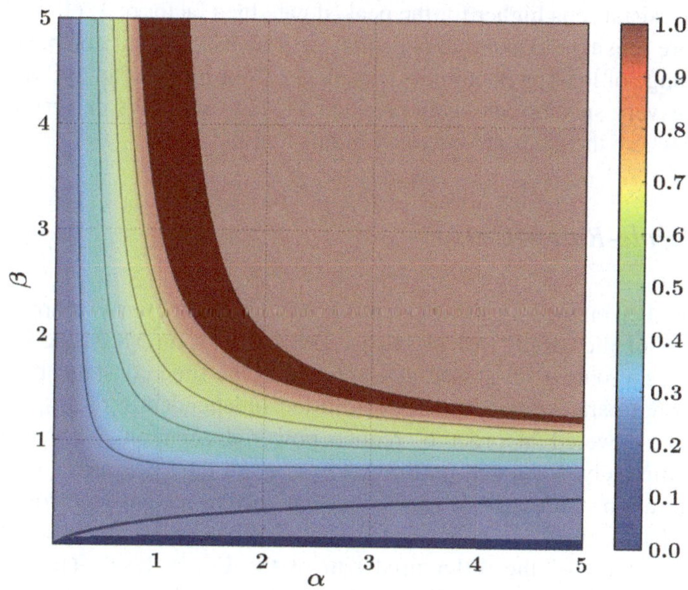

Fig. 3.24 Overshoot ($V_{out,\max}$) in the negative-capacitance peaked MCML system as a function of α and β

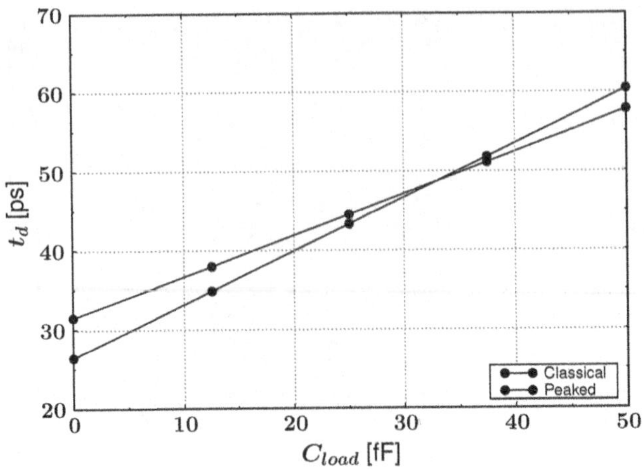

Fig. 3.25 Typical delay versus load curves of classical and peaked MCML gates

the overall power dissipation. In practice, for a total current of I_{SS}, the core MCML gate in a peaked circuit will have a tail current of $I_{SS} - 2 \cdot I_p$, while $2 \cdot I_p$ are allocated to the negative capacitance circuit. At equal power dissipation, a peaked gate can have an intrinsic delay smaller than a classical gate thanks to a reduction of the devices sizes and therefore of the input and output capacitances. However, since the output resistance is higher for the peaked gate by a factor of $1/(1 - 2I_p/I_{SS})$, it will be more sensitive to external parasitics, and its use will only be beneficial over a limited range of loading conditions (Fig. 3.25). Thus, peaking in MCML gates can be used for very specific applications, such as ring oscillators or very high-speed datapaths, where the gates are operated in this region.

3.4.2 *Triple-Rail MCML*

As we have seen previously, the speed of MCML circuits is, to a first approximation, independent of the supply voltage V_{DD}. The delay of an MCML gate is related to V_{sw}/I_{SS}, the output resistance that appears in the time constant; both V_{sw} and I_{SS} are design parameters that can be adjusted independently of V_{DD}. Therefore, V_{DD} can be lowered in order to reduce power dissipation, without any speed penalty. Ultimately, V_{DD} will be limited by the tail current source, which should stay in saturation in order to deliver a constant current and provide common-mode rejection. If fact, the drain voltage of the tail current source is always defined by the source voltage of the differential pair at the lowest level. This voltage can vary during switchings as described in Eq. (2.60), and the minimum value, which is always reached at $V_{id} = 0$ during the switching, depends on the common-mode voltage at the inputs $V_{cm} = V_{DD} - V_{sw}/2$, the threshold voltage, and the size of the transistors which is a function of the voltage swing and noise margin.

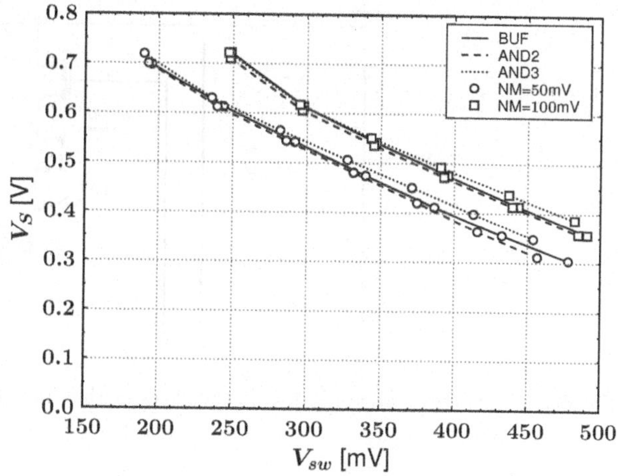

Fig. 3.26 Minimum voltage across the tail current source transistor as a function of NM and V_{sw} at $V_{DD} = 1.2\,\text{V}$ in 90 nm CMOS technology

This voltage is plotted in Fig. 3.26 as a function of the voltage swing, for different noise margin specifications. Depending on the size of the tail current transistors, the minimum drain voltage can be as low as a few hundreds of mV. As it is clear from Fig. 3.26, where the supply voltage is at the nominal value of 1.2 V for the 90 nm CMOS technology, there is an amount of headroom which allows to reduce the supply voltage in order to save power. In practice however, when load devices are implemented with PMOS transistors biased in the linear region, reducing the supply voltage also reduces the maximum gate-source bias voltage for those devices. The consequences of this are twofold:

- with reduced bias voltage, the devices have to be enlarged to keep their equivalent resistance constant. This implies an increase of parasitics at the output nodes, which tends to degrade the intrinsic switching speed of the gates.
- as it was discussed in the previous chapter, the reduction of the gate to source voltage of PMOS devices implies an increased nonlinearity in their characteristics, which in turns implies an increase of the power supply noise.

These effects are the sole consequence of reducing the gate to source voltage of the PMOS load devices. Therefore, if a negative supply voltage is available, this decrease can be avoided [2]; in such a configuration, the gate voltage of the PMOS devices is biased at a voltage lower than ground, as illustrated schematically in Fig. 3.27.

In triple-rail MCML, the supply voltage can be scaled without any impact on the transistor sizes, since the gate to source voltage of PMOS devices is kept constant; in fact, the reduction of the body effect in NMOS transistors even has a positive impact. Sample simulation results are displayed in Fig. 3.28b, where it can clearly be seen that the delay in triple-rail MCML is not affected by the scaling of the supply voltage, leading to a reduction of the power-delay product, while the delay of classical MCML is increasing at reduced V_{DD}.

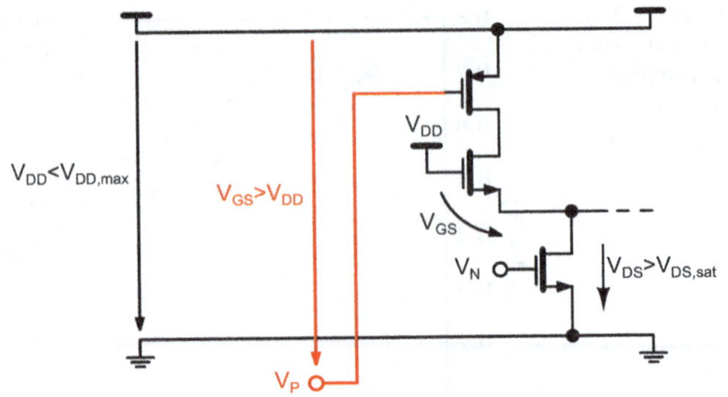

Fig. 3.27 Negative biasing of PMOS load devices in triple-rail MCML

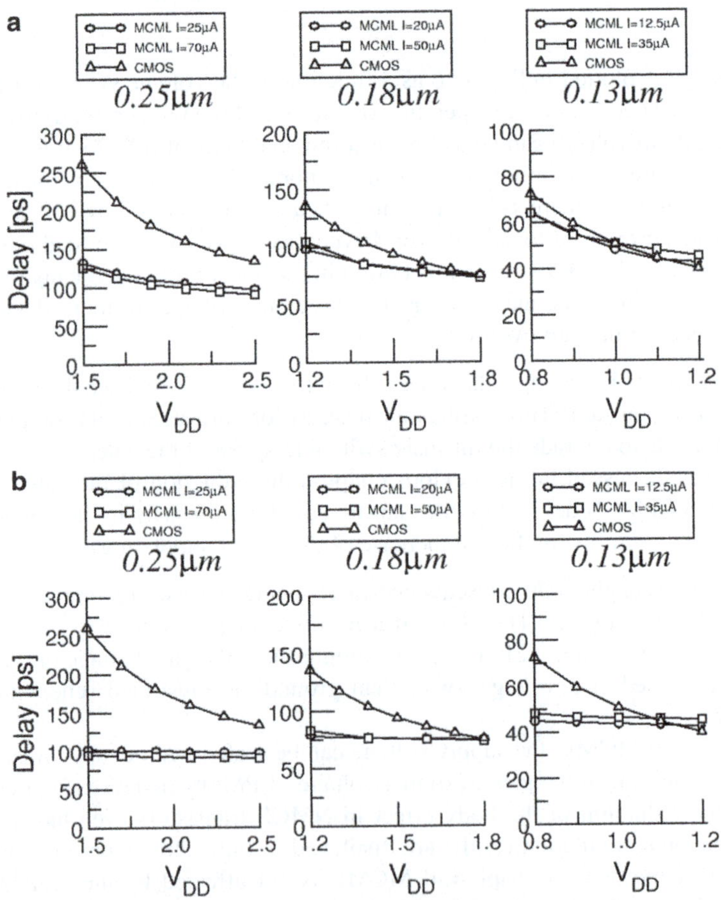

Fig. 3.28 (**a**) Supply voltage scaling in classical MCML. (**b**) Supply voltage scaling in triple-rail MCML

Table 3.2 Power, Delay, and PDP for 15-stages ring oscillators in 0.13 μm technology (TR = triple-rail)

	$V_{DD} = 1.2$ V		$V_{DD} = 0.8$ V		
	MCML	CMOS	MCML	**TR-MCML**	CMOS
Delay (ns)	42.5	40.1	64.2	**45.0**	72.6
Power (μW)	14.9	94.2	10.0	**10.0**	24.8
PDP	636	3776	640	**448**	1799

The bold values identifies the solution with the lowest power-delay-product (PDP)

Table 3.3 Comparison of supply noise (current ripple on the supply line in μA) for 15-stages ring oscillators

	MCML	TR-MCML	CMOS
(a) *Nominal* V_{DD}			
0.25 μ m	0.27	0.27	53.6
0.18 μ m	0.25	0.25	56.8
0.13 μ m	0.03	0.03	29.7
(b) *Minimum* V_{DD}			
0.25 μ m	0.32	0.24	29.5
0.18 μ m	0.35	0.20	35.6
0.13 μ m	0.05	0.03	14.7

Tables 3.2 and 3.3 show results of using triple-rail MCML to lower the supply voltage in different technologies. The results are compared with regular MCML and CMOS.

Triple-rail MCML can achieve this power savings at the cost of an additional negative power supply. It is interesting to notice that this additional power supply has to deliver an extremely low power, since it is used for biasing the gates of PMOS devices, and the DC current is due to gate leakage only (in addition to any power required by the biasing circuitry itself). Such a negative voltage can be generated on chip with high power efficiency and relatively low area cost by using a charge pump circuit, for example.

References

1. S. Badel, Y. Leblebici, An inductorless peaking technique applied to MOS current-mode logic gates, in *IEEE Norchip Conference*, November 2004
2. S. Badel, Y. Leblebici, Breaking the power-delay tradeoff: Design of low-power high-speed MOS current-mode logic circuits operating with reduced supply voltage, in *IEEE International Symposium on Circuits and Systems (ISCAS)* (2007)
3. H.T. Bui, Y. Savaria, 10GHz PLL using active shunt-peaked MCML gates and improved frequency acquisition XOR phase detector in 0.18μm CMOS, in *IEEE International Workshop on System-on-Chip for Real-Time Applications* (2004)
4. H.T. Bui, Y. Savaria, Shunt-peaking in MCML gates and its application in the design of a 20 Gb/s half-rate phase detector, in *IEEE International Symposium on Circuits and Systems (ISCAS)* (2004)
5. R. González Carvajal, J. Ramírez-Angulo, A.J. López-Martín, A. Torralba, J.A. Gómez Galán, A. Carlosena, F. Munoz Chavero, The flipped voltage follower: a useful cell for low-voltage low-power circuit design. IEEE Trans. Circuits Syst. Part I **52**, 1276–1291 (2005)

Part II
Design Automation for Differential Circuits

Chapter 4
Design Methodology for MCML Standard Cells

4.1 Standard Cells and Semi-custom Design

Application-specific integrated circuit (ASIC) design processes are usually classified into either the full-custom or the semi-custom categories. A full-custom approach is one where components in the system are all custom-tailored at transistor-level, and the system is optimized as a whole; this approach can yield very efficient results but the development costs are extremely high for complex circuits. In the semi-custom approach, a set of custom standard cells is created and optimized, and these cells are used as primitive blocks for implementing the system with the aid of specific automation tools. Sometimes, both approaches are combined, with, for example, a full-custom design of critical elements such as very high-speed datapaths, and semi-custom design of less demanding parts of the system.

4.1.1 Semi-custom Flow Overview

In a typical semi-custom design flow, roughly described, the circuit is architected at the behavioral level with a hardware description language (HDL) such as VHDL or Verilog. The resulting circuit behavior is then validated against the specification with RTL simulations. Next, the circuit is mapped to a network of primitive components (netlist) by using a logic synthesis tool, which generates a circuit composed of standard cells whose behavior is equivalent to the HDL description, and optimizes the resulting implementation based on design constraints and characteristics of the standard cells. At this stage, a first estimation of the different performance metrics (power dissipation, circuit speed, silicon area occupation) is available. The next step is the physical implementation of the circuit, using the placement and routing (P&R) tool. The P&R tool usually proceeds in two steps: first, the cells are placed in the available area with the objective of minimizing interconnection length; then, the

S. Badel et al., *Design Automation for Differential MOS Current-Mode Logic Circuits*, https://doi.org/10.1007/978-3-319-91307-0_4

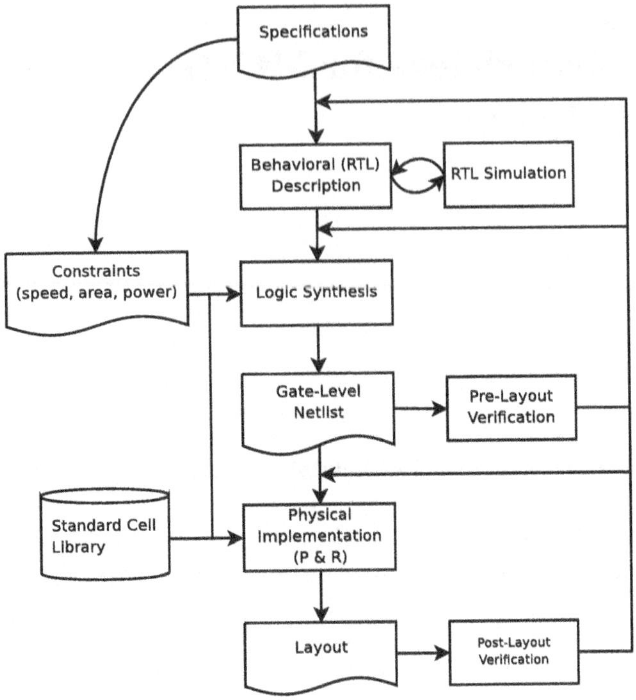

Fig. 4.1 Semi-custom design flow

interconnections are physically routed. After the physical implementation, the final numbers for the performance metrics can be extracted by using accurate (sign-off) extraction of parasitics, timing, and power analysis. If the final numbers meet the specifications, the final layout can be extracted for tape-out; otherwise, turnarounds will be required by reoptimizing the synthesized netlist or maybe even the RTL architecture (Fig. 4.1).

4.1.2 Standard Cells

Physically, a semi-custom circuit is typically composed of standard cells placed in rows, as illustrated in Fig. 4.2. Standard cells are designed to be abutted horizontally, and can have varying widths but all must have the same height.

On the top and bottom of each cell is drawn a rectangle of metal extending from the left edge to right edge of the cell, that eventually connects to neighboring cells, resulting in *power rails* that cross the entire circuit horizontally; these rails are usually connected to *power rings* around the circuit. In order to save area, it is

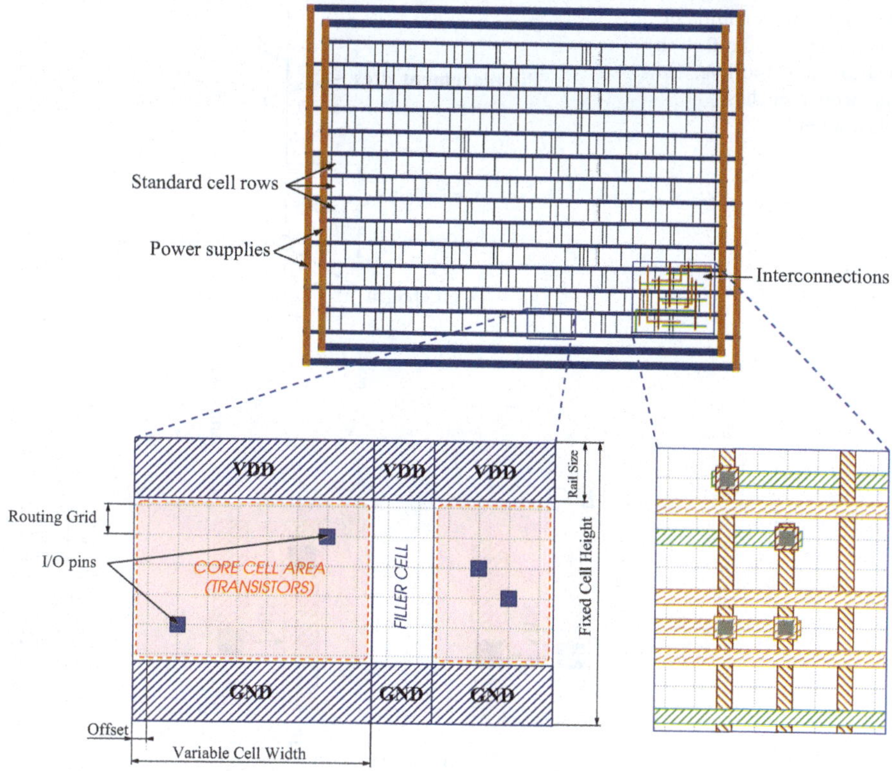

Fig. 4.2 Physical view of a semi-custom circuit and standard cells

common to share the power rail between two adjacent rows, by flipping every other row, resulting in a power-ground-power-ground pattern.

The interconnections are routed on a regular *grid* and connect to *I/O pins* in the standard cells; these pins should be placed on the routing grid to allow easy connection by the router. To guarantee this, *routing grids* are defined for each routing layer, according to the technology design rules, which can be different for different layers. Usually, each routing layer is also assigned a *routing direction,* either horizontal or vertical, which the router will use preferentially in order to maximize the availability of routing tracks. The pitch of the routing grid is typically made as small as possible, in order to maximize the routing resources, and limited by the minimum spacing rule. As shown in Fig. 4.3b, when vias require a larger metal width than the wire width, the pitch can be chosen according to three different situations:

- via-to-via, commonly used, allows two vias to be placed on adjacent tracks without violating design rules
- via-to-line is a tighter routing pitch, which allows a wire to run next to a via to be placed on an adjacent line, but two vias cannot be placed adjacently

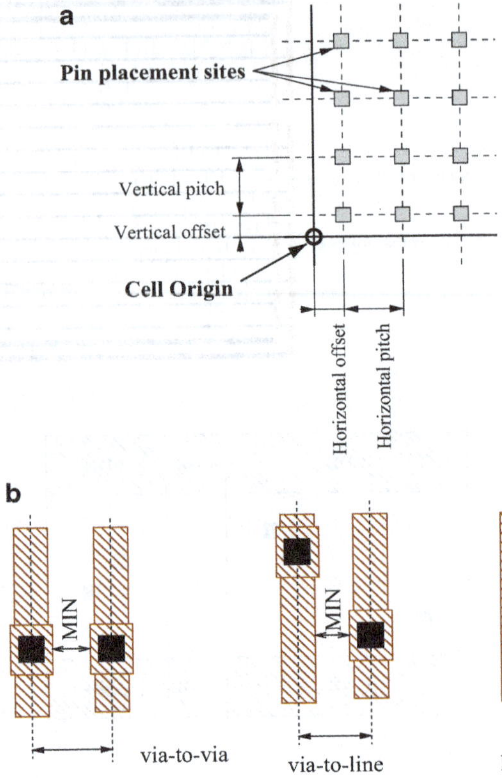

Fig. 4.3 (**a**) Definition of routing pitch and offset. (**b**) Via-to-via, via-to-line, and line-to-line pitches illustration

- line-to-line is the smallest possible pitch, where a wire cannot run next to a via. Practically, this results in every via blocking adjacent routing tracks, and is not commonly used.

Note that modern technologies often have relaxed design rules for vias, which allows the metal wire not to be extended when placing a via; this basically makes all three spacings equal.

Standard cells are also placed on a regular grid—namely the *placement grid*—whose pitch is an integer multiple of the routing grid pitch; this way, pins inside the placed cells will always be located on the routing grid. In order to guarantee the placement of pins for all layers, the placement pitch should be a common multiple of all different routing pitches; practically, standard cell pins are always on the bottom layers (first, sometimes second metal) and the placement grid is typically derived from the grids of the first two routing layer.

To allow the placement and routing engine a maximum of flexibility in optimizing the physical implementation, standard cells should be made so that it is possible to flip them horizontally. In order for the pin sites to remain on the grid when a cell is flipped horizontally, its width should be an integer multiple of the placement grid. Similarly, the height should be a multiple of the placement grid in order to allow flipping the cells vertically.

In order to maximize the number of possible pin placement sites inside the cells, the routing grid can be shifted with respect to the placement grid; this defines the *routing offset* (Fig. 4.3). The amount of shifting should be equal to exactly one half of the pitch in order to preserve the grid when a cell is flipped horizontally.

All of these requirements are as much design parameters as they are constraints on the design (essentially, layout) of standard cells. They have to be chosen carefully in order to maximize the overall performance of the library.

4.2 Logic Gates Synthesis

One of the essential questions that must be addressed in order to develop a library of cells is the synthesis of logic functions—that is, given a logic function to be implemented, how can we construct an MCML circuit that realizes it. This naturally yields to the reverse question: what are all the logic functions that can possibly be realized by MCML logic networks, given that the complexity of the network is limited in some way. With these information, it will be possible to select a number of functions for a standard-cell library and implement them.

Before answering these questions, it is useful to introduce some formalism establishing a bidirectional relationship between physical MCML logic networks and the logic function that they perform. This can be achieved through the notion of binary decision diagrams.

4.2.1 Binary Decision Diagrams

Binary decision diagrams are one of the multiple ways of representing a logic function. Binary decision diagrams are popular because of their compactness and their easy manipulation through graph algorithms, and as we will see, they naturally have a one-to-one correspondence with MCML logic networks, which makes them a tool of choice.

A binary decision diagram is a directed graph representing a logic function, where leaf nodes give the result of the function, each internal node represents one variable and each edge (weighted 0 or 1) represents a value assigned to the variable corresponding to the node from where it is originating. As an example, Fig. 4.4 illustrates the binary decision diagram for the logic function $F(A, B) = \overline{A} + \overline{B}$. In the BDD, any line of the truth table can be evaluated by following the path defined by the edges with weight corresponding to the value of their associated variable in this line of the truth table. For example, in the third line of the truth table, we have $A = 1$ and $B = 0$; starting from the root node (A) in the BDD, following the branch $A = 1$ leads to the B node, then following the branch $B = 0$ leads to the 1 node, which is the value of the function for this set of inputs. A BDD is said to be ordered,

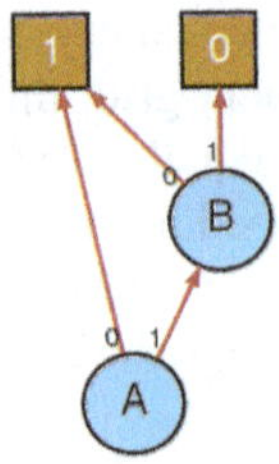

(a)

A	B	$F(A,B)$
0	0	1
0	1	1
1	0	1
1	1	0

(b)

A	B	$F(A,B)$
0	?	1
1	0	1
1	1	0

(c)

Fig. 4.4 Binary decision diagram and truth table for the function $F(A, B) = \overline{A} + \overline{B}$ (**a**) BDD (**b**) truth table (**c**) simplified truth table

if the variables always appear in the same order as one walks the tree from the root to the leaves. An ordered BDD is usually called an OBDD.

In an MCML logic network, the basic processing element is the MOS differential pair, which acts as a current switch. The differential pair receives current from the source node and outputs it to one of the drain nodes, depending on the input voltage. The input voltage is differential, and after defining one polarity representing the logic one state and the opposite polarity defining the logic zero state, and labeling each drain node as '0' or '1' accordingly, the function of the differential pair can be interpreted as a binary function: when the input is one, the current is steered to the '1' node; when it is zero, the current is steered to the '0' node. By applying this methodology to all the differential pairs in an MCML network, one is able to trace the path of the current from the tail current source to either one of the complementary outputs, for each possible set of input values. Defining a polarity for the output as well, we note that when the current is steered to the positive output, it is pulled low and thus logically the value of the output signal is zero; therefore, the positive output can be labeled $\boxed{0}$ indicating that when the traced path leads to this node, the result is a logic '0'—the opposite is true for the negative output node.

As illustrated in Fig. 4.5, the analogy between MCML logic networks is immediate: each differential pair corresponds to a node in the BDD, each interconnection corresponds to an edge, and the output nodes correspond to the '0' and '1' leaf nodes of the BDD. The root of the BDD—the node which has no incoming edge—corresponds to the bottom-most differential pair in the MCML network, which has its source connected to the drain of the current source. Thus, there is a one-to-one correspondence between MCML logic networks and binary decision diagrams; moreover, the BDD can capture not only the logical function of the network, but also its physical properties: the number of transistors and their interconnections. Therefore, in order to produce an MCML network that realizes a given function, one can generate the corresponding BDD and map it to a physical network; additionally, optimization of MCML networks can be realized by manipulation of the corresponding BDD.

Fig. 4.5 Analogy between a BDD and the corresponding MCML logic network

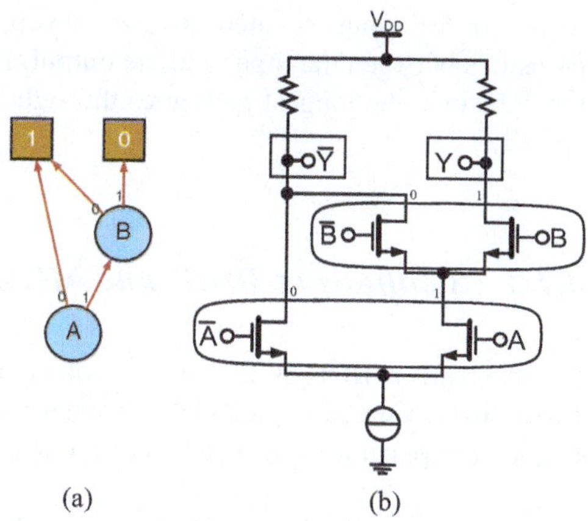

4.2.2 *Analysis of BDDs and MCML Networks*

The analysis of a BDD is the task of extracting information from a given structure. The extracted information can be related to the logic function performed by the BDD and its properties, and in the present context, it can pertain to properties of the equivalent MCML network.

Given a BDD, as we have seen, one can evaluate the result of the logic function for a given set of input values by tracing the path corresponding to the set of input variable values, starting from the root and ending at either one of the leaf nodes. Therefore, each path in the BDD starting from the root and ending at the $\boxed{1}$ leaf defines an implicant of the function—that is, a product term implying a true value of the function. Thus, by tracing all paths starting from the root and ending at the $\boxed{1}$ leaf, it is possible to find all implicants, defining the on-set of the function; the function can then be written as the sum of these implicants. Conversely, writing the sum of cubes corresponding to paths ending at the $\boxed{0}$ results in an expression of the complemented function.

As an example, considering the BDD of Fig. 4.5a, there are two paths from the root to the $\boxed{1}$ node: $\overline{A} \rightarrow \boxed{1}$ and $A \rightarrow \overline{B} \rightarrow \boxed{1}$. The function is thus $F(A, B) = \overline{A} + A\overline{B} = \overline{A} + \overline{B}$. Conversely, there is a single path to the $\boxed{0}$ node, which is $A \rightarrow B \rightarrow \boxed{0}$, and the inverted function can be written as $\overline{F(A, B)} = AB$.

Regarding the properties of the corresponding MCML network, as we have seen, each node corresponds physically to a differential pair. The number of levels of the network, i.e. the number of stacked differential pairs, is given by the length of the *longest path*. As it was seen in Chap. 2, this number has a strong influence on the performance of the gate: the higher the number of levels, the larger all transistors must be sized to preserve the noise margin, therefore the larger the speed and area penalties. Also, for a given node in the network, its position is given by the longest path starting from this node. As it was seen, the delay from a given input to output increases for inputs located at lower levels in the network, due to the increased parasitics between the input and the output. In the BDD of Fig. 4.5, the A input is at level 2, since the longest path goes through B and two edges, and the B input is at level 1.

4.2.3 Synthesis of BDDs and MCML Networks

Synthesizing a BDD is the process of creating a BDD that implements a given function. Given a function, a BDD can easily be constructed by cofactor expansion. Cofactor expansion is the process of recursively applying the identity

$$\begin{aligned} F(x_0,\ldots,x_i,\ldots,x_n) &= x_i \cdot F(x_0,\ldots,1,\ldots,x_n) + \overline{x_i} \cdot F(x_0,\ldots,0,\ldots,x_n) \\ &= x_i \cdot F_{x_i} + \overline{x_i} \cdot F_{\overline{x_i}} \end{aligned}$$

where F_{x_i} and $F_{\overline{x_i}}$ are the *cofactors* of F with respect to x_i, until eventually the resulting cofactors are all equal to either 1 or 0. In the BDD, one such operation corresponds to creating one node x_i, expanding the cofactors F_{x_i} and $F_{\overline{x_i}}$ into two subtrees, and creating two edges weighted '1' and '0' from the x_i node to the root node of each of these subtrees. Indeed, taking any node x_i in a BDD, this node is the root of a subtree of the entire BDD, which is a BDD in itself and represents a boolean function G. Evaluating the function for $x_i = 0$ or 1 means following one of the outgoing edges, each leading to a new node which is the root of a subtree; these two subtrees by definition represent the value of G for $x_i = 0$ and $x_i = 1$, respectively, which are the cofactors of G with respect to x_i, proving the validity of this method of construction. Note that the cofactor expansion naturally yields an OBDD if the variables are completely expanded in order.

As an example, the expansion of the function $F(A, B) = \overline{A} + \overline{B}$ from Fig. 4.5b into cofactors and the construction of the BDD is illustrated in Table 4.1. Note that the third step is unnecessary in this particular case: since $F_{\overline{A}} = 1$, an edge could have been drawn directly between A and $\boxed{1}$, resulting in the BDD of Fig. 4.5a. Nevertheless, it is included for the sake of clarity in illustrating the process of synthesizing the BDD. Besides, this highlights the fact that two different BDDs can be structurally different but logically equivalent. In the coming paragraphs, we will show how to reduce a BDD to a minimal and canonical form.

4.2.4 Reduction of BDDs

When a BDD is synthesized by fully expanding the cofactors in all variables in order, an n-input function will always result in a BDD with $2^N - 1$ nodes. This is easily deduced by considering that the first variable will generate the root

Table 4.1 Synthesis of BDD by cofactor expansion

Step	Expansion	BDD
F	$F(A,B) = A\cdot\overline{B}+\overline{A}\cdot 1$ $= A\cdot F_A+\overline{A}\cdot F_{\overline{A}}$	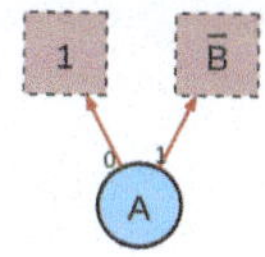
F_A	$F_A = \overline{B}$ $= B\cdot 0+\overline{B}\cdot 1$	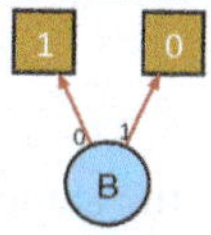
$F_{\overline{A}}$	$F_{\overline{A}} = 1$ $= B\cdot 1+\overline{B}\cdot 1$	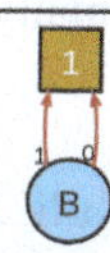
3	$F(A,B) = A\cdot\left[B\cdot 0+\overline{B}\cdot 1\right]+\overline{A}\cdot 1$	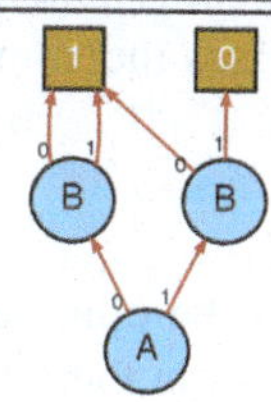

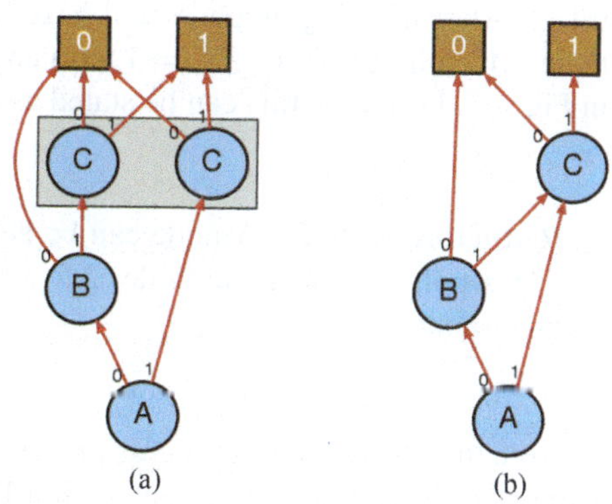

Fig. 4.6 BDD reduction by merging isomorphic subgraphs

node, and that subsequent variables will each generate two subtrees. Clearly, this representation is far from optimal in most of the cases, because the resulting tree is redundant.

Redundancy is present when the BDD contains unnecessary nodes or parts that are identical. It corresponds to terms that can be eliminated or factored out. For example, consider the function $F(A, B, C) = (A + B) \cdot C$. Cofactor expansions in the order $A - B - C$ result in the BDD of Fig. 4.6a. Note that C was factored out in the initial expression of F; in the BDD, there are two C nodes whose outgoing

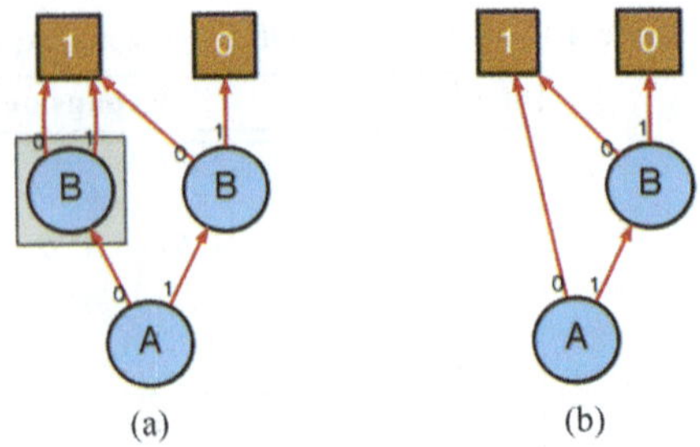

Fig. 4.7 BDD reduction by eliminating trivial factors. (**a**) before reduction. (**b**) after reduction

edges point to the same nodes: they are redundant. They can be merged as illustrated in Fig. 4.6b. Formally, a rule can be formulated as

Reduction Rule 1 Two nodes can be merged in a BDD if they are the root of equivalent (isomorphic) subgraphs. One of the subgraphs is then deleted, and all edges incoming to its root node are redirected to the root of the other subgraph.

Terms can be eliminated when the function is independent of a variable. This gives cofactors in the form of $x_i \cdot F + \overline{x_i} \cdot F$ which can clearly be simplified to F. Since in the BDD, both subgraphs of the nodes are identical, they can be merged by rule 1. After merging, node x_i will have both outgoing edges pointing to the same node, which implies a $x_i + \overline{x_i} = 1$ term and the node can be eliminated as illustrated in Fig. 4.7. Formally, this can be stated as a rule

Reduction Rule 2 A node can be deleted in a BDD if both of its outgoing edges point to the same node. All incoming branches are redirected to the latter node.

Note that the reduction process preserves the order of variables, hence an OBDD remains ordered after reduction; such a BDD is called a *reduced ordered BDD* or ROBDD. In an ROBDD, though the variables appear in the same order for all paths, it may happen that some paths do not contain all the variables.

4.2.5 Variable Ordering and Optimum Implementation

Interestingly, it can be shown that, when recursively applying rules 1 and 2 to reduce an ordered BDD until no further reduction is possible, the OBDD is reduced to a *canonical* form. This means that two OBDDs with the same variable ordering will

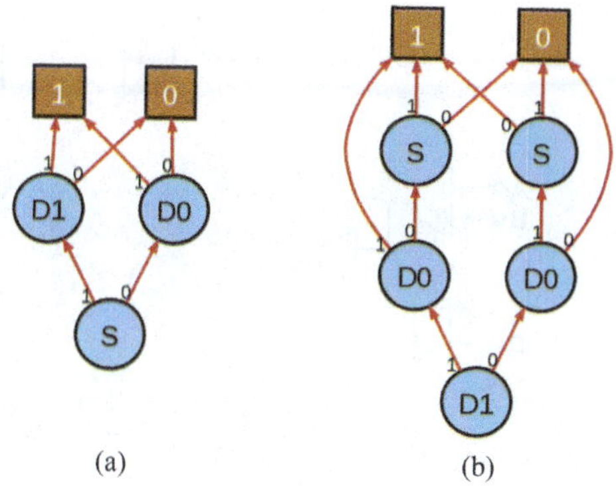

Fig. 4.8 Influence on the variable ordering on the result of cofactor expansion of the MUX2 function. (**a**) $S - D_0 - D_1$ ordering. (**b**) $D_1 - D_0 - S$ ordering

always be identical after reduction. Note that this canonical form depends on the initial ordering of the variables: the variables expanded first will always be located at lower levels in the resulting OBDD. In fact, different variable orderings can produce very different results. For example, consider the 2-to-1 multiplexer function $F(D_1, D_0, S) = S \cdot D_1 + \overline{S} \cdot D_0$: two different results of cofactor expansion are illustrated in Fig. 4.8, with the expansion performed in the order $S - D_0 - D_1$ and $D_0 - D_1 - S$. In the first case, the resulting BDD has only two levels and three nodes; in the second case, it has three levels and five nodes.

From a logical point of view, these two BDDs are equivalent, i.e. they represent the same function, and the BDD in (a) is obviously more compact than the BDD in (b). From an electrical point of view, however, the two networks present different characteristics. Specifically, the network in (b) will most likely result in a smaller delay from input S to output, since that input is located on the highest level of the network, while inputs D_0 and D_1 will be slower than in network (a). The input capacitances, on the other hand, will be lower in network (a) since there is only one differential pair per input, and the transistors are smaller. In terms of area, the (a) network is obviously advantageous. Nevertheless, if a gate with a low S-to-output delay is desired, implementation (b) may be preferred; this highlights the fact that the compactness of the network can not be a criterion to rule out variants implementing the same function, and, depending on the requirements, any input ordering may result in an implementation which is "optimal."

These considerations suggest that, in order to find an optimal implementation for a given function, it is sufficient to synthesize and reduce the OBDDs for all possible variable orderings, which yields $N!$ possibilities, and to select the most efficient. However, it should not be forgotten that the process of building a BDD is recursive. Starting with a selected variable x_i, one will first expand the function F into cofactors with respect to x_i as $F = x_i \cdot F_{x_i} + \overline{x_i} \cdot F_{\overline{x_i}}$. Next, a second variable will be chosen to expand F_{x_i} and $F_{\overline{x_i}}$. In order to yield an ordered BDD, both should be expanded with the same variable ordering; however, if it is not necessary to obtain an ordered BDD, F_{x_i} and $F_{\overline{x_i}}$ can be expanded with different orderings. Clearly, F_{x_i} and $F_{\overline{x_i}}$ cannot always be synthesized optimally with the

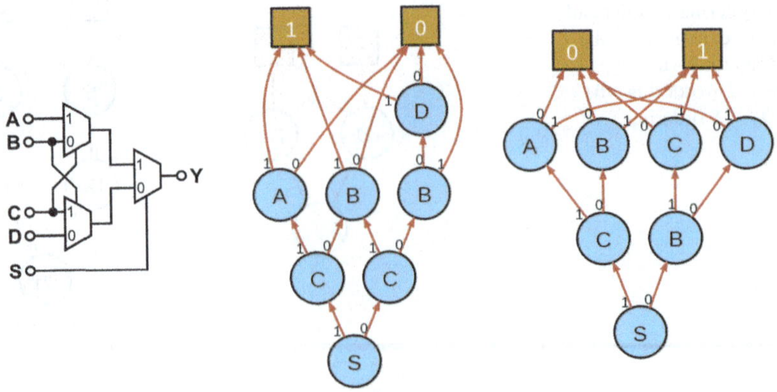

Fig. 4.9 Optimum implementation with free variable ordering

same variable ordering; therefore, suppressing the ordering restriction may yield more efficient implementations. This is illustrated in Fig. 4.9, with the expansion of the function $F(A, B, C, D, S) = S \cdot [C \cdot A + \overline{C} \cdot B] + \overline{S} \cdot [B \cdot C + \overline{B} \cdot D]$. Schematically, this function can be represented with multiplexers as depicted in Fig. 4.9a. Clearly, if S is expanded first, both subtrees will be multiplexers and, as seen with the example of Fig. 4.8, the multiplexer implementation is sensitive on the variable ordering. Therefore, if we constrain the variable orderings to be the same for both subtrees, they can not be realized both with a two-level implementation, as illustrated in Fig. 4.9b, resulting in a four-level implementation for F. If we relax this constraint, we can implement both subtrees with two levels only, resulting in a more compact, three-level implementation as shown in Fig. 4.9c.

When relaxing the ordering constraint, the resulting BDD might not be *ordered*, in the sense that the definition given previously implies (i.e., that the variables appear in the same order for *any* path). A BDD with free variable ordering is called a *free BDD* or FBDD. An FBDD can be reduced according to the same rules as presented previously for OBDDs. Note that the reduction of an FBDD will be canonical too, since an FBDD can always be partitioned into a set of OBDDs, which reduce to canonical forms; only, there are many more possible variable orderings in an FBDD than in an OBDD. In an FBDD, the number of possible variable orderings can be deduced by recursion as follows.

Let us denote by x_N the number of variable orderings for a function of N variables. Let us then partition the graph by separating the root node and its two child subgraphs. The root node can be any of the N variables, and each subgraph can use a different ordering of the $N - 1$ remaining variables. Therefore, there are $(x_{N-1})^2$ possible ways of combining different orderings for the two subgraphs, and we can write $x_N = N \cdot (x_{N-1})^2$. Obviously, we have $x_1 = 1$, and by recursion one can calculate x_N for any value of N.

Usually, a constraint is imposed in the definition of the FBDD, that no path can contain the same variable more than once. This constraint is automatically satisfied when building the BDD from cofactor expansion, because after one variable is expanded, the resulting cofactors are independent of that variable and there is no reason to expand this variable anymore. In the general case, if a path contains the same variable more than once, then all paths going through the '1' edge of the second x_i node will contain terms in the form of $x_i \cdot x_i$, while paths going through the '0' edge will contain terms in the form of $x_i \cdot \overline{x_i}$. In the first case, the term reduces to x_i, and in the second case it is not satisfiable. Therefore, the second x_i node can be deleted according to the following rule

Reduction Rule 3 In a BDD, when a path contains two nodes with the same variable, then the second node (when tracing the path from the root) can be deleted. Edges incoming to that node are redirected to the node pointed to by its active edge ('1' or '0', depending on the active polarity of the variable in this path), and the subgraph rooted at the node pointed to by its inactive edge is deleted.

In practice, given an arbitrary BDD, care should be taken when deleting the subgraph if part of it is shared by other paths which do not pass through the deleted redundant node. The subgraph can be processed by recursion as follows: delete the root and its two outgoing edges, then for each subgraph, repeat only if the root does not have any incoming edges anymore.

4.2.6 Multi-Stage Decomposition

When implementing BDDs into MCML networks, it can become impractical when the BDDs have too many levels, since each additional level results in an overall increase of the delays and transistor sizes in the resulting network. Besides, there is a practical limit in the maximum number of levels dictated by supply voltage requirements.

When a BDD exceeds the practical maximum number of levels, it can be decomposed into multiple stages. The decomposition involves selecting subtrees in the original BDD, and taking them out as separate networks. Each of these subtrees will be replaced by a single node, whose variable value is the result of the subtree it replaces. The subtrees should not have outgoing edges, except at the highest level, and all outgoing edges should end at no more than two different nodes. Similarly, the subtree should only have incoming edges at the root node.

This is illustrated in Fig. 4.10 where a three-level network is decomposed into two two-level networks. Note that this means that physically, the function will

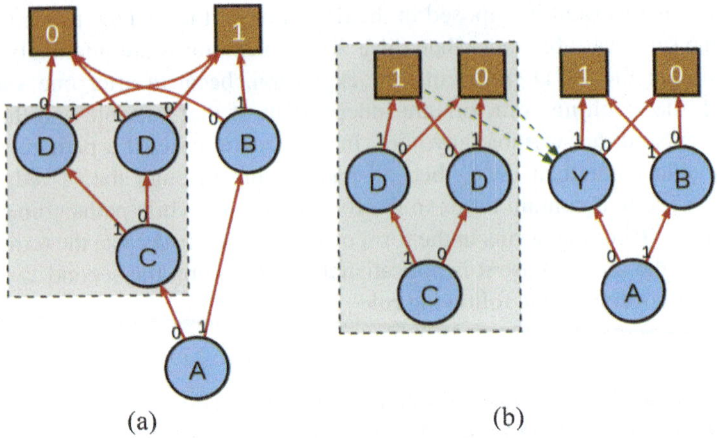

Fig. 4.10 Decomposition of BDDs into multiple stages. (**a**) before reduction. (**b**) after reduction

be implemented using multiple gates; this implies a multiplication of the power dissipation, and additional gate delays.

4.3 Template Approach for MCML Standard-Cell Library

In order to allow efficient optimization by the synthesis tools, a standard-cell library should include a large variety of functions, and these functions should be implemented as efficiently as possible.

In the previous section, it was shown how to produce MCML networks that perform given functions, and how to produce efficient implementation by testing with different variable orderings. An approach for building a standard-cell library could thus involve selection of a number of function to implement, and then searching for and implementing efficient MCML gates for those functions.

However, as it was highlighted, searching for optimal implementations is a time-consuming task, especially with free variable ordering. As an example, searching for an optimal implementation of a 4-to-1 multiplexer, which is a function of six inputs, involves testing more than 10^{13} possible BDDs, which is clearly not tractable. Moreover, given a function, there is no guarantee that this process will result in a feasible implementation.

Selecting a subset of functions which is known to be implementable efficiently results in a suboptimal utilization of the capabilities offered by the logic style. In fact, certain classes of functions can be realized very efficiently in MCML, but these could easily be overlooked since they are "exotic" for the mainstream CMOS logic style. In CMOS logic style, efficient implementations are found for functions written as sum of products, where variables appear either complemented or not complemented, but not both. Each term is implemented as NMOS switches connected in series, and all series switches are connected in parallel; the PMOS network is complementary to the NMOS network. A CMOS cell library essentially contains gates realizing NAND and NOR, And-OR-Invert, and Or-And-Invert efficiently; Exclusive-NOR and Multiplexer are commonly found but much less efficient since they require inverting one or more variables.

In contrast, MCML cells can handle functions which require both inverted and non-inverted inputs efficiently, since the inversion of differential signals is costless. Multiplexers and Exclusive-OR are easily realized with a single stage of logic. However, an MCML logic network cannot have fewer levels than the maximum number of variables in one of its implicants, which limits the efficient realizations to classes of functions where product terms contain only a limited number of variables.

Given that the efficiency of logic function implementations is very different in MCML and CMOS, it would probably not be effective to select a typical set of functions and implement them. Instead, the reverse approach can be taken: searching for functions that can be implemented efficiently, granted that the search space is limited; this is the idea underlying the template approach.

4.3.1 MCML Footprints

In MCML circuits, a practical limitation is given by the number of levels in the logic network, as the performance degrades quickly when increasing the number of levels, due to the larger transistor sizes that are required to keep the noise margin large enough, and to keep the total voltage drop below the minimum required for the tail current source to stay saturated under worst-case supply voltage conditions. Therefore, when searching for implementable functions, the search space can be limited to functions that can be realized with a network of at most N levels. The maximum number of levels essentially depends on the supply voltage and threshold voltage of the transistors. In Fig. 2.26, simulation results are plotted for AND gates with 2 up to 4 inputs; as it can be seen, the delay of the slowest input in the AND4 gates is larger than the equivalent delay of two AND2 gates. Such a gate would only very rarely be used by a synthesizer. Based on these considerations, we limit ourselves to networks of up to three levels in the rest of this work, which already cover a wide range of functions. Nevertheless, the methods presented are valid in the general case.

The most complex network than can be built with N levels is a network where each node has two children which are not children of any other node (except for the leaf nodes). Such a network has exactly $2^N - 1$ nodes; therefore, it cannot realize any function of more than $2^N - 1$ inputs, which gives a limitation of the search space. However, with three levels this results in functions of up to 7 inputs, which is still an unpractical number.

In order to further reduce the search space, we can search for all possible BDD topologies that have at most N levels. Consider the following process of building a N-level BDD:

1. create the leaf nodes $\boxed{0}$ and $\boxed{1}$
2. create a string of N nodes, and connect them with one edge between each pair of adjacent nodes.
3. for the topmost node, create two outgoing edges; these edges must point to the leaf nodes, otherwise the network would have one additional level. Moreover, each edge must point to a different node, or it would result in a subgraph that can be reduced by rule 2. Therefore, the top node has two edges, one pointing to $\boxed{0}$ and one pointing to $\boxed{1}$.
4. process the highest remaining node that has less than two outgoing edges. if this node is at the highest level, proceed as step 3. Otherwise, for each edge to be created, either direct it to an existing node at a higher level (including leaf nodes) or direct it to a new node created at any level above the current level.
5. repeat step 4 until all nodes are processed

Note that in this process, whenever a node is created, it is assigned a unique variable name. However, the weights are assigned randomly; in fact, inverting the weight of the edges is always possible by inverting the variable of the node from which they are originating.

Exploring all possible outcomes of this algorithm is a tractable problem for practical number of levels, and will result in all possible N-level BDDs topologies, that we will call *footprints* hereafter. For networks of one up to three levels, it results in 19 unique footprints. Each of these footprints corresponds to a different physical network.

Each N-level footprint has a number of nodes comprised between N and $2^N - 1$. In the search process, we have assigned a different variable name to each of the nodes, corresponding to the function with the maximum number of inputs that can be realized with this network. Functions with less inputs (down to N) can be realized by assigning some variables to more than one node. By testing all possible variable assignments, it is possible to deduce all functions that can be realized by a particular footprint. Assigning a set of inputs to the nodes of a footprint will result in one particular specialization of this footprint; we will call such a specialization a *template*. In doing this, many redundant functions will be encountered, which should be ruled out. Functions can be redundant because they are equivalent to another function by permutation or inversion of the inputs or outputs. In order to identify them, it is useful to introduce some notions on the equivalency and classification of boolean functions.

4.3.2 Classification of Boolean Functions

A boolean function of N variables $F(x_1, x_2, \ldots, x_N)$ is a function from the set $\{0, 1\}^N$ to the set $\{0, 1\}$, i.e. for every set of input values, where each input can be either 0 or 1, the function associates a value of either 0 or 1. There are 2^N possible combinations of input values, and for each of these the function can associate one of two values, therefore there are 2^{2^N} different functions of N variables.

Amongst these functions of N variables, some will be independent of one or more of the variables; i.e. they will also be included one or more sets of functions of M variables, with $M < N$. The number of functions that involve N and exactly N variables can be calculated by subtracting all functions of less than N variables from the number 2^{2^N}, which by recursion leads to $\sum_{i=0}^{N} (-1)^N \cdot C_i^N \cdot 2^{2^{N-i}}$ where C_i^j denotes a binomial coefficient.

This number grows rapidly with N. The number of function of 1 up to 7 variables are indicated in the table below

N	# of functions	Functions
0	2	$0 \quad 1$
1	2	$A \quad \overline{A}$
2	10	$AB \quad \overline{A}B \quad A\overline{B} \quad \overline{AB}$ $A+B \quad \overline{A}+B \quad A+\overline{B} \quad \overline{A}+\overline{B}$ $A \oplus B \quad \overline{A \oplus B}$
3	218	...
4	64,594	...
5	$4.2946 \cdot 10^9$	...
6	$1.8447 \cdot 10^{19}$	...

Amongst the 2^{2^N} functions of N variables, some functions can be classified as equivalent according to a given transformation. Two functions are equivalent under a transformation if one can be turned into the other by applying that transformation. The set of functions that are equivalent under a certain transformation form an equivalence class.

In logic applications, one important transformation is the permutation of inputs. Two functions are said to be *P-equivalent* if one can be transformed into the other by a permutation of its inputs. In most of the cases, and specifically in the case of standard-cell design, the choice of freely assigning the inputs of a gate is always available. Therefore, it is redundant to implement more than a single function from the same set of P-equivalent functions.

An extension to this transformation is to allow the negation of inputs and output. Two functions are said to be *NPN-equivalent* if one can be transformed into the other by a permutation of its inputs, and the negation of zero or more inputs or output. In MCML logic circuits, because the signals are differential, inputs and outputs can be freely inverted. Therefore, any gate can not only implement a set of P-equivalent functions, but more generally a set of NPN-equivalent functions.

Therefore, in MCML circuits, it is sufficient to implement one function from each class of NPN-equivalent functions of N variables to be able to realize all functions of N variables with a single gate. Formulas are given in [2] to calculate the number of P- and NPN-classes of functions of N variables, these numbers are summarized in the table below

N	P-classes	NPN-classes
2	8	2
3	68	10
4	3904	208
5	37,329,264	615,904
6	25,626,412,300,941,060	200,253,951,911,058

4.3.3 MCML Templates

Previously, we have derived the possible unique implementations of MCML networks with up to N levels, which we call footprints. We stated that an N-level footprint has a number of nodes M comprised between N and $2^N - 1$, and that by connecting multiple inputs together, a set of functions of N up to M variables can be realized. In the following, we will establish a procedure to derive those functions, and by applying it to all footprints we will derive the complete, non-redundant set of functions implementable with MCML logic networks up to N levels.

Because a single MCML network is able to realize a complete NPN-class of functions, we shall extract only a single function per class. We will call the resulting gate, produced by assigning specific inputs to a footprint, a *template*. Each template generates an NPN-class of function by permutation of inputs and inversion of inputs/output.

Let us consider an N-level footprint with M nodes. In order to produce a function of P variables, $P \epsilon [N, M]$, a set of inputs $\{x_1, x_2, \ldots, x_P\}$ has to be mapped onto the M nodes of the network. Moreover, each node can receive a variable either complemented or not, and the output can be inverted. Labelling the M nodes as $\{a_0, a_1, \ldots, a_M\}$, this defines a function from $\{x_i\}_{i=1 \ldots P}$ to $\{a_i\}_{i=1 \ldots M}$, with the following constraints:

1. each node should be assigned a variable
2. each variable should be mapped to at least one node
3. no variable may appear more than once in a path of the BDD

Moreover, we wish to identify all functions that are NPN-equivalent in this process.

Considering the assignment of the variables or their complements to the nodes of the BDD, there are M nodes and P variables which can be complemented, therefore $(2P)^M$ possible assignments. From this number, we must deduce the assignments that contain less than P variables. There are C^P_{P-1} ways of choosing a set of $P - 1$ variables, therefore the number of assignments μ^M_P of exactly P variables on a M nodes network satisfies $\mu^M_P = (2P)^M - C^P_{P-1} \cdot \mu^M_{P-1}$. This results in $\mu_P = \sum_{i=1}^{P} (-1)^{P-i} \cdot C^P_i \cdot (2 \cdot i)^M$.

Table 4.2 Number of possible variable assignments of P variables on an M-nodes network with N levels

N	M	P	ν_P^M
1	1	1	1
2	2	2	1
	3	2	12
		3	1
3	3	3	1
	4	3	20
		4	1
	5	3	260
		4	30
		5	1
	6	3	2800
		4	560
		5	42
		6	1
	7	3	27,216
		4	8400
		5	1064
		6	56
		7	1

Amongst these assignments, we wish to rule out assignments that are equivalent by permutation of variables. Since there are $P!$ ways of permuting P variables, we should divide the result by $P!$ Moreover, we also wish to rule out all assignments that are equivalent by inversion of variables. There are 2^P possible inversions of zero or more of the variables, therefore the final number of assignments that should be tested is

$$\nu_P^M = \frac{\sum_{i=1}^{P}(-1)^{P-i} \cdot C_i^P \cdot (2 \cdot i)^M}{2^P \cdot P!} \tag{4.1}$$

Values of ν_P^M are summarized in Table 4.2 for MCML footprints with one up to three levels. This number is much smaller than $(2P)^M$, allowing to considerably reduce the search space. When applying these input assignment to a footprint, it can be verified that rule 3 is satisfied. If rule 3 is satisfied, then the resulting function is validated, as well as the function resulting from inverting the output. In this process, functions will still be generated that are P- or NPN-equivalent due to symmetries in the network. In order to identify these functions, we need to test them for equivalence with previously validated functions.

P- and NPN-equivalence can be tested very easily using the concept of signature. A signature is defined as a value that can be calculated, that is characteristic of a certain property. An example of defining a signature for a boolean function is to write its truth table, then interpreting the right-most column (which gives the function result) as a binary number. A function of N variables having a truth table of 2^N lines, this results in a binary number with 2^N digits which is different for each function. This number can be calculated and stored, then for each function to be tested, its signature can be compared to the stored signature. In order to check for P- or NPN-equivalence, one can calculate the signatures for all P- or NPN-variations of the function and store them; then, to test a function for equivalence, its signature is calculated and compared to all the P- or NPN-signature collected.

The usage of signatures for equivalency testing is not very efficient, and other methods exist such as the one described in [1], nevertheless it is a simple approach that is sufficient for testing the functions that we are considering here for networks of up to three levels.

4.3.4 *Proposed Set of Standard Cells*

Applying the methodology described previously for MCML networks of up to three levels results in 19 unique footprints, which implement 108 templates. Each template implements a class of NPN-equivalent functions of 1 up to 7 inputs. The 19 footprints are presented in Figs. 4.11, 4.12, and 4.13, each accompanied by a few representative templates that they implement. For a detailed description of the 108 templates, please refer to Appendix B.

Note that, amongst those 108 templates, only 65 are unique from a logical point of view. The remaining 43 implement functions already existing in the set of unique functions, but are physically implemented with a different network, and hence have

Fig. 4.11 One-level templates

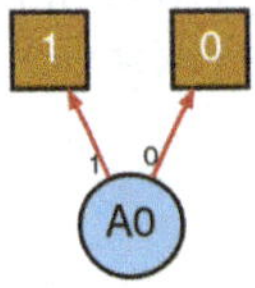

F1
BUF

Fig. 4.12 Two-level templates

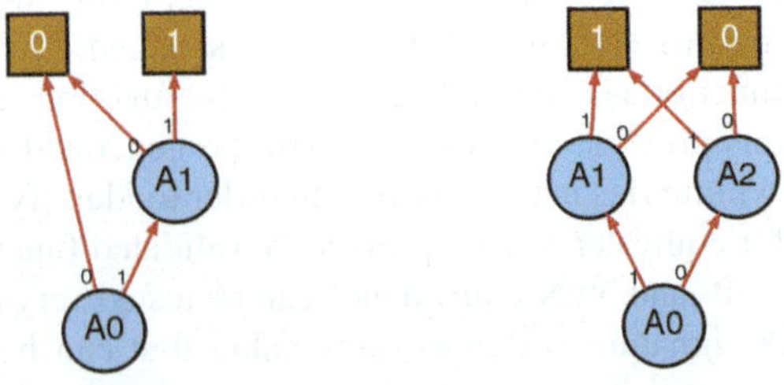

F2
AND2

F3
XOR2, MUX2V1

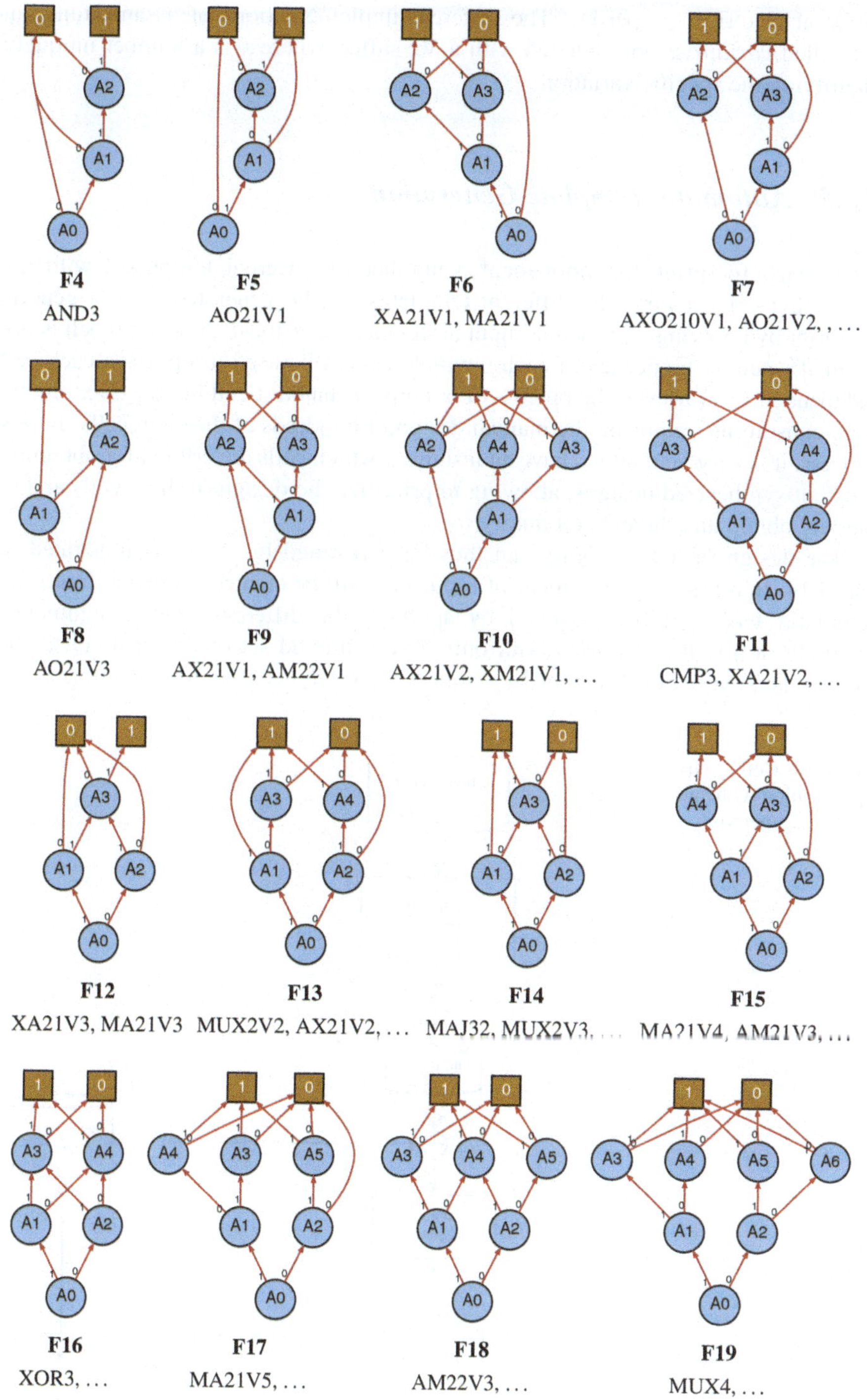

Fig. 4.13 Three-levels templates

different electrical properties. The different implementations of a same function are called *variations* and denoted with a Vx suffix, where x is a number uniquely identifying the specific variation.

4.3.5 Automatic Template Generation

Since each footprint can implement a number of different templates, with the same physical network, the different templates can be generated from a generic footprint by applying a particular input assignment. Obviously, this approach is not optimal: specific properties of each template cannot be used to optimize each cell independently. However, the process of building a standard-cell library is extremely time-consuming, requiring the manual design of hundreds of different cells. In this process, it is very valuable to have an insight of which cells are critical for obtaining efficient synthesized designs, allowing to prioritize the design of these cells and to quickly obtain an efficient cell library.

The design of a cell library can thus be performed by steps, as illustrated in Fig. 4.14. In a first step, a number of footprints will be created. For each footprint, templates will then be generated by applying the different input assignments, resulting in a rich cell library with only a very limited set of physical cells. The library can then be benchmarked by synthesizing reference designs, and information

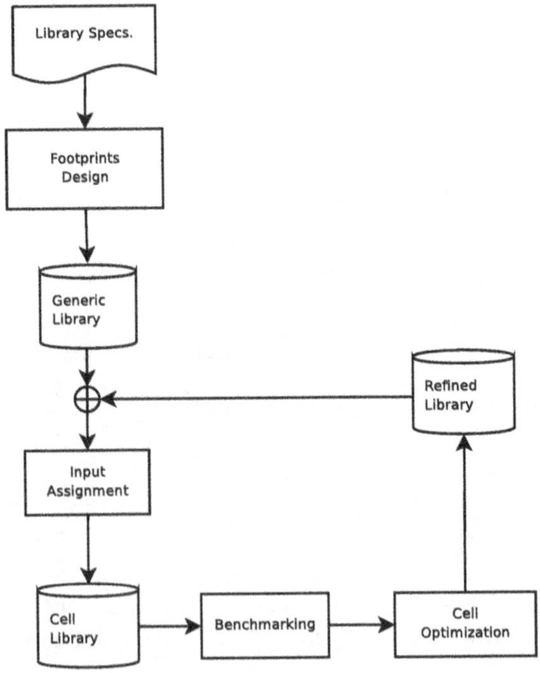

Fig. 4.14 Cell library optimization with automatic template generation

can be extracted regarding the frequency of utilization of each cell. In further steps, sets of cells can be selected to be optimized, resulting in a new, refined library, including optimized cells complemented by automatically generated templates.

4.4 Standard-Cell Design

It is clear that the final performance of semi-custom designs will heavily depend on the quality of the cell library. Ideally, standard cells should be fast, allowing to satisfy demanding timing requirements, dissipate little power and require as few silicon area as possible, and at the same time be robust and immune to process, temperature and supply voltage variations in order to provide a high yield, a long lifetime and operate reliably. In practice, some of these objectives are somehow convergent, but most are contradictory; all have weighted importance that are very application-specific, and formulating generic design rules that encompass all objectives is unfortunately not possible.

4.4.1 Design Parameters

Semi-custom designs are usually characterized by a high density of cells and interconnects. In modern technologies, interconnects account for a large part of the parasitics, and it is not sufficient to optimize cells independently, without considering their environment. In fact, the physical size of cells in a library will have a direct impact on the overall wire length; therefore, when cells are sized up, the increase in drive strength will be counterbalanced by an increase in parasitics due to the increased wire lengths.

Typically, and this is true in particular for MCML, each cell has an intrinsic delay due to its internal parasitics that is independent of the cell size, because the internal parasitics grow proportionally to the drive strength. In terms of area, transistor sizes are proportional to the drive strength as well, and cell area increases accordingly. The drive strength of a cell is characterized by the inverse of the slope of its delay-load characteristic, which is its equivalent output resistance, and the delay of a cell for a given load condition can be expressed as

$$t_d = t_{d,i} \cdot \left(1 + \frac{C_L}{C_i}\right) \quad \text{where } t_{d,i} = R_o \cdot C_i \tag{4.2}$$

where C_i represents the internal parasitics of the gate.

When cells are sized up, the ratio of C_L over C_i is decreased, and cells operate at a speed which is closer to their intrinsic speed limit. However as we said, sizing up cells also implies an increase of C_L. The load capacitance is composed of two parts: C_{FO}, the parasitics due to the input capacitances of driven gates (fanout), and C_w, the parasitics due to wiring. Fanout load increases proportionally to the drive strength, and when this component is dominating the overall load, sizing up cells hardly has any effect on the resulting circuit speed and results in waste of power. On the other hand, wire length does not increase proportionally to the area of a circuit, but rather at the pace of the half perimeter of the circuit, i.e. proportionally to the square root of the area. Therefore, when wiring dominates the overall load, sizing up cells allows to reduce the ratio C_L/C_i, and thus to increase circuit speed.

Therefore, we can write an expression for the delay as a function of the drive strength as

$$t_d(D) = t_{d,i} \cdot \left(1 + k_{\mathrm{FO}} + \frac{k_a}{\sqrt{D}}\right) \tag{4.3}$$

where D denotes the drive strength, k_{FO} and k_a are factors characterizing cell input capacitance and area per unit of drive strength, respectively. The power dissipation in MCML cells is proportional to their drive strength,

$$P(D) = k_P \cdot D \tag{4.4}$$

The efficiency of a cell family can be characterized by the different k factors, and by the intrinsic speed of the cells. Cells with low intrinsic delay k_{FO} and k_a will require lower drive strength to operate at a given speed, while cells with low k_P will dissipate less power at a given drive strength.

In MCML circuits, k_P is proportional to the voltage swing, since cells with lower voltage swings have lower output resistance. However, lowering the voltage swing implies increasing transistor sizes at a rate approximately proportional to $(V_{sw})^{-2}$, as it was seen in Chap. 2. This affects not only the area of the cells—and thus the wire load—and their input capacitances, but also the intrinsic delays. Therefore, cells with higher voltage swings will be more performant in standard-cell applications.

4.4.2 Cell Layout

As it was highlighted throughout this discussion, designing compact cells is critical to achieve high speeds as well as low power dissipation. In MCML circuits, each cell is composed of three parts: the load devices, the tail current source, and the logic network. A cell library will be composed of cells that all have the same height, containing very different logic networks. Naturally, load devices can be placed close to the power rail, and current sources close to the ground rail, leaving space in the middle area for laying out the logic network (Fig. 4.15). MCML cells need two

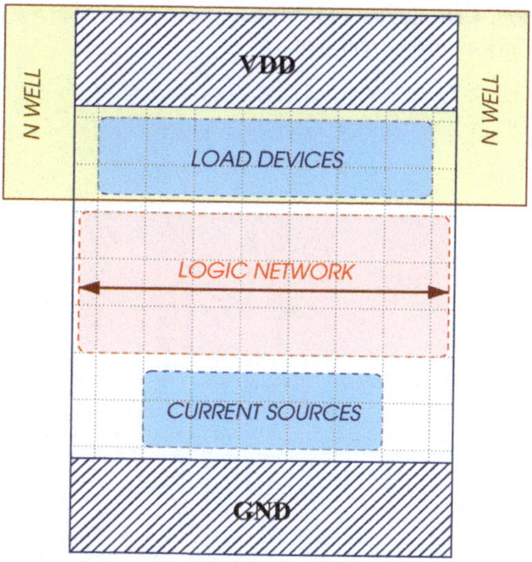

Fig. 4.15 Generic layout template of an MCML standard cell

additional rail for the bias voltages of load devices and current sources. These rails can be placed at the top and bottom edges of the cells, and accessed via polysilicon routing under the power and ground rails. In this way, a single metal layer is needed and the access to power and ground rails from the core with first metal is possible.

The load devices, as well as the current sources, should be composed of identical unit devices connected in parallel, in order to ensure a good matching. Therefore, the size of these areas will be identical for all cells sharing the same drive strength. However, different cells will have different logic network to implement, whose size can range from just a single pair of transistor to 14 transistors for the most complex three-level network. When the cell width has to be increased to accommodate a larger network, area stays unoccupied in the top and bottom of the cell.

A challenging aspect of MCML standard cell layout is to produce efficient layouts for this wide variety of cells. There are several practices that help ensuring a uniform layout for all types of cells, as far as this is possible:

- Select identical horizontal size for the unit load and current source devices. This way, when devices are repeated in parallel to produce higher drive strengths, the overall size of the load and current source areas will remain identical, avoiding unused silicon area.
- Lay out load devices in two rows, and current sources in a single row. Since each gate contains two times more load devices than current sources, this will result in equal width for both areas.
- For cells with a small logic network, when free area in the middle part of the cell allows it, load devices and current sources can be folded into multiple rows to save area (Fig. 4.16).

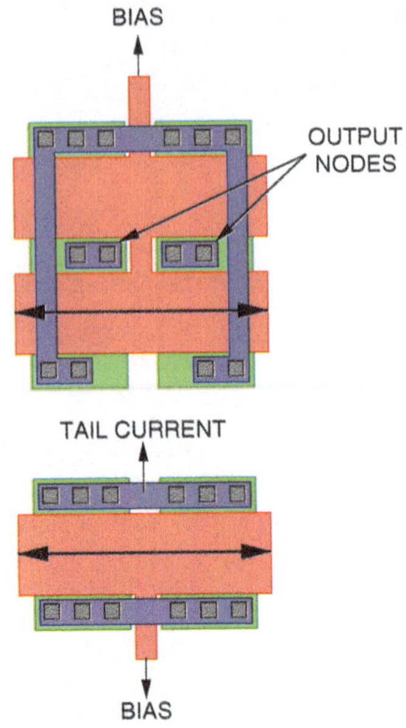

Fig. 4.16 Layout of load devices and current sources

4.4.3 Unit Cell Sizing

The size of the unit cell is an important parameter. A unit cell is a cell whose tail current is provided by a single transistor, as well as each of the two load devices. Cells with higher drive strength will all have tail currents that are integer multiple of the unit tail current.

Practically, a unit cell should be as small as possible, and the minimum size is essentially limited by design rules and matching requirements. The unit current source and load device should be sized several times larger than the minimum, both in width and length, in order to provide a reasonable matching. According to the selected values, the tail current should be chosen as large as possible, in order to increase the drive strength/area ratio. Then, the logic networks of the different cells can be sized, and the cell height chosen accordingly.

References

1. A. Abdollahi, Canonical form based boolean matching and symmetry detection in logic synthesis and verification. Ph.D. thesis, University of Southern California, 2006
2. M.A. Harrison, Combinatorial problems in boolean algebras and applications to the theory of switching. Ph.D. thesis, University of Michigan, College of Engineering, 1963

Chapter 5
Design Automation for Differential Circuits

5.1 Overview

When design flow was briefly introduced in the previous chapter, RTL simulation, synthesis, and P&R tools were mentioned. In fact, many more are required, each dealing with a specific aspect of the design process, and a more complete list could be

- Cell library characterization
- Event-driven simulation
  - register-transfer level (RTL)
  - gate-level with parasitics back-annotation (post-layout, SDF)
- Logic synthesis
- Static timing analysis (STA)
- Noise analysis
- Placement & Routing
- Layout parasitics extraction
- Power grid analysis
- Transistor-level simulation

Currently, design automation tools are tailored for the usage of CMOS cell libraries, and the adoption of a different logic style is often problematic due to the lack of specific design automation tools.

In particular, with MCML circuits, most automation tools fail to accommodate to the differential nature of the cells. In order to successfully use existing tools, a number of problems must be solved, that will be discussed in this chapter. The approach that will be described involve the usage of two different "views" of signals in the circuit. Signals in MCML circuits are differential, which means that the information is carried in the difference of two floating voltages. Thus, there is only a single signal, physically carried by two complementary voltages.

S. Badel et al., *Design Automation for Differential MOS Current-Mode Logic Circuits*, https://doi.org/10.1007/978-3-319-91307-0_5

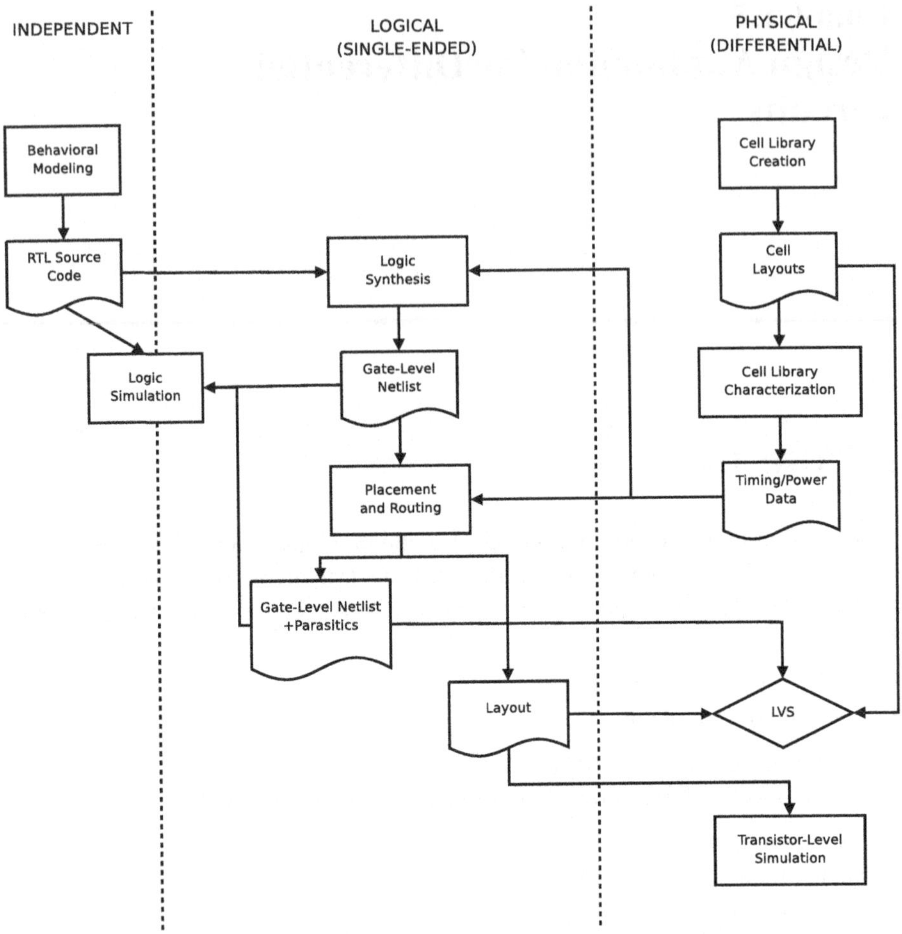

Fig. 5.1 Design flow for differential circuits

When considering a circuit from a logical point of view, the knowledge that physically one signal has two complementary supports is unnecessary; therefore, from a logical point of view, all signals can be reduced to the information that they carry, and represented as a single entity. However, when dealing with physical tasks, it is obviously necessary to take it into account. Thus it will be necessary to switch between the two views of the circuit in design process, depending on the task to be executed. This is illustrated in Fig. 5.1 as a flowchart, with dotted lines representing the boundary between the different realms: physical, logical, and independent. Each arrow crossing the boundary between logical and physical areas requires to modify the view of the circuit.

5.2 Logic Synthesis

During logic synthesis, a behavioral description of a circuit is translated into a netlist of cells picked from a target library. From a functional point of view, all a synthesizer needs to know about the cells is what function they perform. Synthesis tools also perform optimizations of the timing, area, and power dissipation based on user-defined constraints; for this task they require detailed information about the cells. The extraction of these information is performed during library characterization; in this section, we will focus on issues specific to synthesis of MCML circuits.

5.2.1 *Synthesis with Differential Cells*

From a logical point of view, a library of MCML cells is no different from any other library: it contains a set of cells which perform certain functions, and circuits are built by interconnecting cells according to their functional description. However, the cells operate on differential signals, and one immediate question is how the synthesizer can handle this differential nature.

In the synthesis process, the cells are considered as functional units, i.e. processing elements that produce output signals from input signals. Making known to a synthesizer details about the physical representation of those signals is unnecessary, and it is only natural to merge the two polarities of a differential pair into a single logical signal (Fig. 5.2). This raises one issue: differential signals can be inverted freely, by inverting the two polarities, but a synthesizer has no knowledge of this.

In order to let the synthesizer invert signals freely, a possible approach could simply be to include inverters in the cell library. However, this approach is not efficient since it will result in an unnecessary waste of power and area, in addition to inserting additional delays in the signal paths. To avoid that, the inverters could be removed after synthesis, but this would invalidate all timing paths through the inverter and jeopardize the timing integrity of the resulting circuit.

A better approach is to provide the synthesizer with different variant of each cell, where inputs and output are inverted in all possible combinations. In this way, the synthesizer will not have to explicitly invert signals, since it will always be able to select a gate with inverted output, or one or more inverted inputs, depending on the needs. The drawback of this approach is that it dramatically increases the number of cells in the synthesis library, since for a cell with N input and M outputs, there are 2^{N+M} possible combinations of inverting the inputs and outputs.

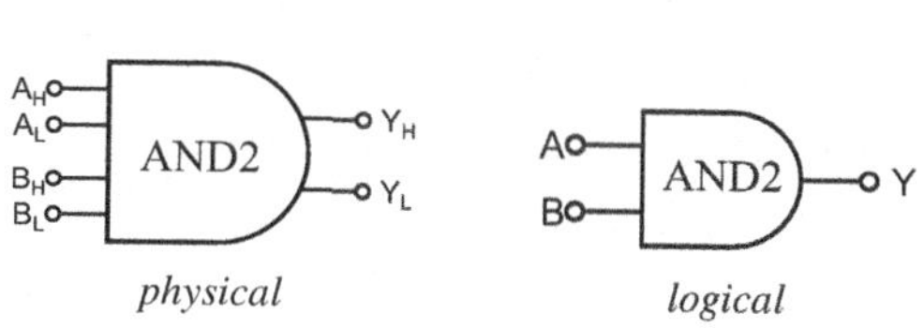

Fig. 5.2 Physical and logical views of an AND2 cell

This does not require to create more physical cells; indeed, the variants of a same cell with inverted pins should all be replaced by the same physical cell during physical implementation, and only the connections to both polarities of the differential pins should be modified as required by the particular variant.

Such a variant can be thought of as a higher-level cell, which encapsulates the master cell and the connection to its pins. Indeed, this is particularly helpful, since it can be applied throughout all the different steps of the flow, ensuring consistency. For example, a variant of a cell can be described in verilog structurally by instantiating a master cell and rearranging the connections. Thus, there is absolutely no post-processing to do on the synthesized netlist before simulating it. Similarly, transistor-level netlists can be generated in the same fashion for each variant, as well as P&R abstracts, allowing a complete consistency and ease of use throughout the complete flow. As we shall see later on, the resulting netlist can also be used for placement and routing without any modification.

When generating variants for the different cells in the library, it is necessary to assign to each a meaningful name, that identifies the unique signal assignment which it implements. Not only does this allow to immediately identify the function of each cell, but this enforces the consistency throughout all the different views of the cell library through the definition of a naming convention. Any naming convention that uniquely identifies an assignment is valid for this purpose; in this work, we propose to sort all the pins of a cell according to their direction and alphabetical order, and to assign to each a bit number in order. In the resulting bitmask, each bit is set to 1 if the corresponding pin is inverted, and to 0 otherwise. The resulting binary number can be translated to a decimal number, which is then appended to the master cell name.

As an example, the AND2 cell of Fig. 5.3 has two input pins A and B, and one output pin Y. Sorting these three pins alphabetically, then placing the outputs first, followed by the inputs results in the bitmask Y, B, A. Therefore, A corresponds to bit 0, B to bit, 1 and Y to bit 2. A variant with input A and output Y inverted results in the bitmask 101, and the variant is thus named AND2_5 (Figs. 5.4 and 5.5).

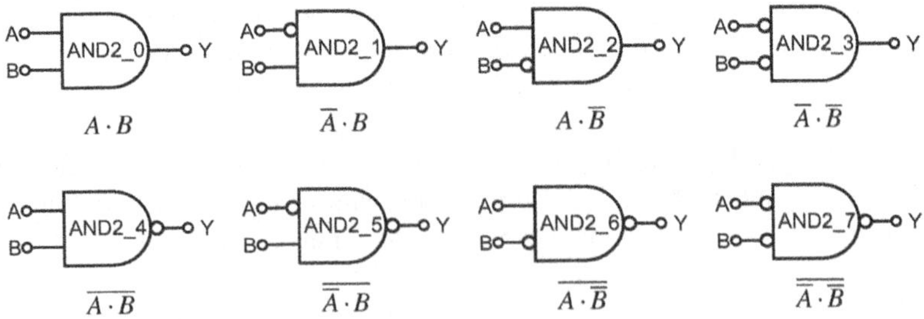

Fig. 5.3 The eight variants of an AND2 cell

```
// 2-Input AND
// Master cell
module AND2(Y,A,B);
output Y;
input A,B;
and I0 (Y,A,B);
endmodule
```

```
// AND2 with inverted A
module AND2_1(Y,A,B);
output Y;
input A,B;
not I0 (NA,A);
AND2 I1 (Y,NA,B);
endmodule
```

Fig. 5.4 Description of a logical variant as a high-level cell: schematic view and verilog HDL description

```
.SUBCKT AND2
 +Y_H Y_L A_H A_L B_H B_L
 * ...
 * Transistor-level netlist
 * ...
.ENDS AND2
```

```
.SUBCKT AND2_1
 +Y_H Y_L A_H A_L B_H B_L
 * Structural netlist
 X0 Y_H Y_L
 + A_L A_H B_H B_L AND2
.ENDS AND2_1
```

Fig. 5.5 Description of a physical variant as a high-level cell: schematic view and SPICE netlist

5.2.2 Bias Generator and Level Converters in the Synthesis Process

Another specificity of MCML circuit is the fact that they require a bias generator block. This block should physically be included in the circuit in order to provide as good a matching as possible, allowing to cancel process variations.

Additionally, in some cases the differential signals used by MCML circuits are not suitable to interface with external circuitry. This can be the case if an MCML block is integrated along with conventional CMOS logic. Also, when signals need to be received or driven off-chip, it is not convenient to interface differential signals that require twice the number of bonding pads, whose number is strictly limited by area limitations and bonding capability. Interfacing signals requires including level

```
## find bias circuit master cell
set master [get_lib_cell -q "BIAS_CIRCUIT"]
set num_pins [get_attribute -q $master number_of_pins]
## create instance
set bias_master [get_attribute $master name]
create_cell $bias_inst $master
set_dont_touch [get_cells $bias_inst] true
## create and connect pins on toplevel
foreach_in_collection pin [get_lib_pins -of $master] {
 set port_name [get_attribute $pin name]
 set port_dir [get_attribute $pin pin_direction]
 create_port -direction $port_dir $port_name
 create_net $port_name
 connect_net [get_net $port_name] [get_port $port_name]
 connect_net [get_net $port_name] [get_pin $bias_inst/
     $port_name]
 set_dont_touch [get_net $port_name] true
}
```

Fig. 5.6 Design compiler script to add bias circuit block

converters at the circuit boundary, that transform the signals to and from MCML-compatible differential voltages.

The inclusion of a bias generator and level converters will require area and contribute additional power dissipation. Additionally, in the case of level converters, it is important that the synthesizer includes them in the timing analysis in order to ensure timing integrity. For these reasons, these cells should be inserted into the circuit before the synthesis process.

In one approach, these cells can be explicitly inserted in the RTL description of the circuit, as this is possible in Verilog as well as VHDL. However, this makes the RTL code technology-dependent, which designers usually try to avoid. In another approach, they can be inserted during the synthesis process, before the actual synthesis process takes place. For example, in *the design compiler* synthesizer, a few commands to be inserted in the synthesis script are sufficient to automatically insert the technology-specific cells. Example scripts are given in Figs. 5.6 and 5.7. In this way, the process is transparent to the user, and the bias generator and the level converters are taken into account during synthesis, allowing maximum accuracy.

5.3 Placement and Routing

After synthesis, the resulting netlist will be forwarded to the place and route tool. In the approach described above, the synthesized netlist is composed of library cells which are logical (i.e., with single-ended pins) variants of a master cell.

```
## find all toplevel input ports
set ports [get_ports [all_inputs]]

##  find level converter cell
set master [get_lib_cell -q "LEVEL_CONV_IN"]
set num_pins [get_attribute -q $master number_of_pins]

## search pins in converter cell master
set iso_master  [get_attribute $master name]
set iso_out_pin [get_attribute [get_lib_pins -of $master
    -f "pin_direction == out"] name]
set iso_in_pin  [get_attribute [get_lib_pins -of $master
    -f "pin_direction == in"]  name]

## process all ports
foreach_in_collection port $ports {

 ## process all nets to be isolated
 set net_to_isolate [get_nets -of_objects $port]
 foreach_in_collection net $net_to_isolate {

  ## create new objects
  set net_name [get_attribute $net name]
  set iso_inst "[regsub {(.*)\[([\d]+)\]$} [get_attribute
    $net name] {\1_\2}]_conv"
  set iso_net  [regsub {(.*)(\[[\d]+\])?$} [get_attribute
    $net name] {\1_d\2}]
  create_cell $iso_inst $master
  set_size_only [get_cell $iso_inst] true
  create_net $iso_net

  ## connect objects
  set connections [remove_from_collection [all_connected
    $net] $port]
  disconnect_net $net $connections
  set direction [get_attribute -q $port port_direction]
  if { $direction == "out" } {
   connect_net    $net         $iso_inst/$iso_out_pin
   connect_net    $iso_net    $iso_inst/$iso_in_pin
  } else {
   connect_net    $net         $iso_inst/$iso_in_pin
   connect_net    $iso_net    $iso_inst/$iso_out_pin
  }
  connect_net      $iso_net $connections
  set_dont_touch [get_net $net] true
 }
}
```

Fig. 5.7 Design compiler script to add level converter cells at the each input port. The script can be modified to process output ports

After placement and routing, the resulting circuit must contain physical cells and physical (i.e., differential) interconnections. The two main issues to be solved are thus:

- routing of differential nets
- replacement of variant cells by their master with correct input/output connections

5.3.1 Routing of Differential Nets

Any P&R tool can process differential signals natively; as far as the tool is concerned, both signals in a pair are just two independent signals, which connect to pins inside standard cells. Therefore, by converting the logical synthesized netlist into a physical equivalent (replacing each signal by a corresponding pair), a differential circuit can be placed and routed as is.

However, there are several problems associated with this approach. First, it multiplies the number of nets to be processed by a factor of 2, which results in larger runtimes. Second, since the router has no knowledge of the relationship between the two complementary signals in a pair, they are both routed independently and can have very different resulting physical implementations. This is detrimental to the signal integrity, since both the noise robustness and the low noise generation features of differential signals rely on the fact that they are physically located close to each other.

The third issue is related to the timing analysis of the circuit. During the whole process of placement and routing, the circuit timing is analyzed and the results drive the tool into generating a physical circuit that satisfies the user-defined timing constraints. In MCML circuits, switching events are triggered by differential voltages at the inputs, but each output node operates its transition independently. Therefore, in the general case, the two output nodes of a gate can have different delays if their loading is different, but both delays are relative to the zero-crossing point of the differential input signal. A complete timing model of an MCML gate should define the output differential signal as a function of the input differential signal, and the individual loading conditions of each output node.

Yet, timing analysis tools can only deal with single-ended circuits, where delays are defined from one signal to another, and it is not possible to express a delay as a function of a differential signal. Timing analysis can be correct, however, with logical cells where each signal represents one differential voltage, if we can make the assumption that both output nodes of each cell are loaded equally. This requires to work with logical cells during the place and route flow on the one hand, and to make sure all signal pairs are routed as bundles, so that their parasitics are closely related.

Executing placement and routing with logical cells implies that each physical pair of signals will be represented by a single wire. Similarly, a logical abstract view should be provided for each cell, where each pair of pins is represented by a

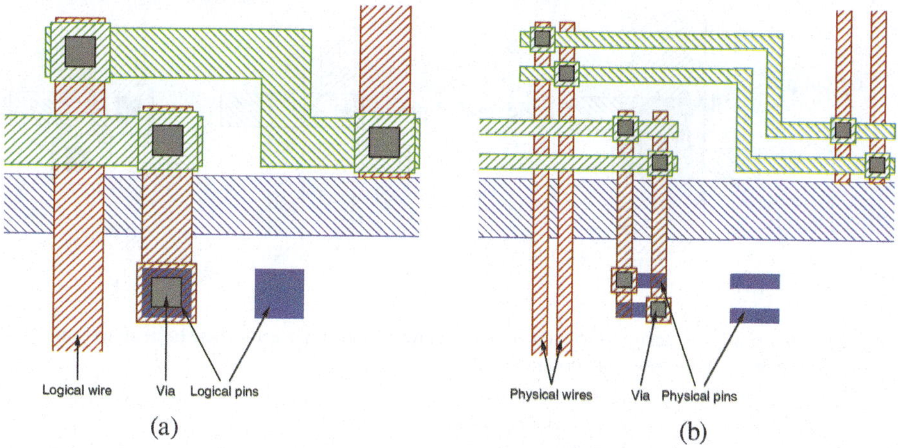

Fig. 5.8 Differential signals routing. (**a**) Routing of logical design (**b**) routing converted to physical

single logical pin. The router will then be able to process the logical design, and in a next step, the logical wires can be split into pairs of wires, as depicted in Fig. 5.8. Similarly, cells will be replaced by their physical counterpart, turning each logical pin into a pair of physical pins.

The connections at the standard cell pins deserve special attention. After a logical wire is split into a pair, each resulting wire must be assigned a polarity. As we have seen previously, according to the synthesis flow there will never be any logical inversion, i.e. each pin will be connected only to pins of same polarity in other cells. However, no convention can be defined regarding the position of specific pin polarities in the cells, therefore it is not possible to know a priori to which of two physical pins a specific wire in pair will be connected to. Therefore, pins should be designed in a way that guarantees that connection from both wires to both pins is possible. A possible way to achieve this is to draw physical pins as metal stripes, in a direction perpendicular to the routing direction of the next routing layer; in this way, the two connections are always possible as shown in Fig. 5.9.

A dedicated tool could handle the logical-to-physical wire transformation as well as the positioning of vias, producing a correct physical design from the routed logical circuit. In a simpler approach, vias can be removed during the wire splitting step, and the physical design undergoes another routing pass to complete the connections.

In the logical routing step, the characteristics of the wires and vias should be defined according to the desired result after splitting. The physical parameters include the wire width, routing pitch, and different quantities defining the via geometry, as depicted in Fig. 5.10. According to these parameters, the corresponding values for logical routing can be deduced.

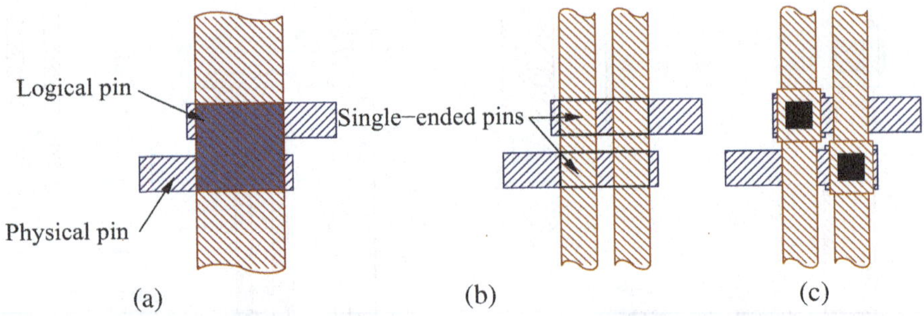

Fig. 5.9 Physical pins design allowing the connection to both polarities (**a**) logical and physical pins placement (**b** and **c**) possible connections

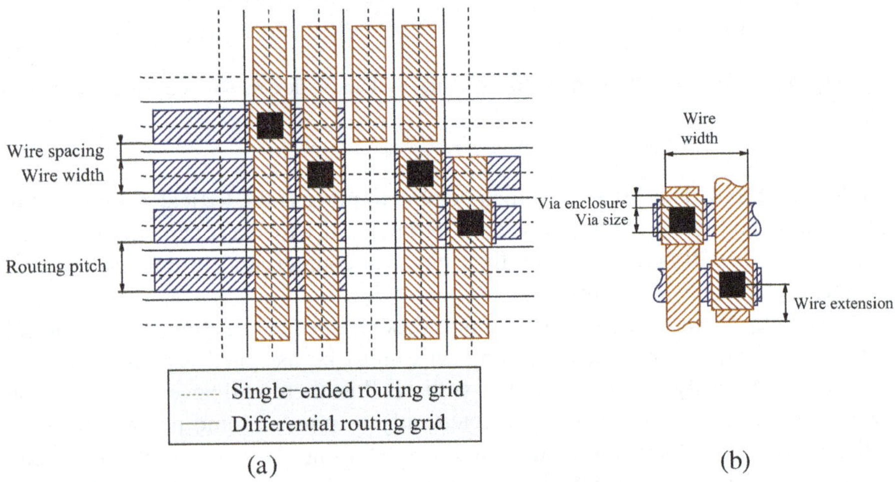

Fig. 5.10 Physical wire and via parameters. (**a**) wire parameters. (**b**) via parameters

In the most general case, the spacing between the two wires in a differential pair and the spacing between two neighboring pairs could be different. However, if the physical design is to be routed again to complete the connection as it was proposed, it is important that a common grid be defined. In this way, wires after splitting will fall on a regular grid, and can be handled efficiently by the router. This compels the two spacings mentioned above to be equal (or, at least, multiples). In this case, logical routing parameters are calculated from the physical parameters as shown in Table 5.1.

Table 5.1 Physical and logical routing parameters

Quantity	Physical	Logical
Routing parameters		
Wire width	W	$W' = 2 \cdot P + W$
Routing pitch	P	$P' = P$
Routing offset	O	$O' = O + \frac{P}{2}$
Via parameters		
Via size	S	$S' = S + P$
Via enclosure	E	$E' = E$
Wire extension	X	$X' = X + P$

5.3.2 Variant Cells in the Place and Route Flow

In the place and route flow as described until here, the variant cells used for synthesis have at no point been replaced by their master cell. This replacement could take place just before place and route, by processing the synthesized netlist to replace the cells and correct the connections. However, this additional step can be avoided by providing abstract views of the variant cells for use by the router.

In this approach, abstract views are created for each master cell. Then, abstracts are generated for the variants by swapping polarities of the pins, just as it is done to generate variants for synthesis. In doing this, the physical connections to the pins will be automatically relocated, as illustrated in Fig. 5.11. When the abstracts are defined in LEF format, the reference to the master cell can be explicitly made by specifying it in a `FOREIGN` statement. This statement specifies which cell should be referenced in the physical library (typically, GDSII) when exporting the layout from the place and route tool. This way, the process is completely transparent to the user.

5.3.3 Parasitics Modeling

During the place and route process, parasitics are extracted from the layout in order to provide accurate data for the timing analysis. Typically, P&R tools rely on fast, approximate extraction based on the wire geometries and capacitance lookup tables. In the flow as proposed above, the wire geometries used by the router are different from the physical wire geometries, therefore it is necessary to adapt the capacitance tables accordingly.

As an example, the capacitance model used in the *Encounter* place and route tool is illustrated in Fig. 5.12. Three types of capacitances are defined:

- area capacitance, corresponding to the coupling between the horizontal side of a wire and the neighboring routing layer
- coupling capacitance, corresponding to the coupling between vertical sides of neighboring wires on the same layer

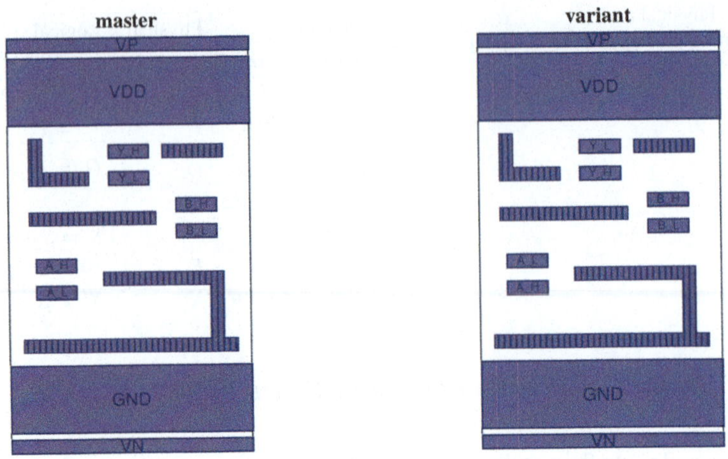


```
MACRO AND2
FOREIGN AND2 0 0 ;
...
PIN Y_H
 # pin definition
END Y_H
PIN Y_L
 # pin definition
END Y_L
...
PIN A_H
 # pin definition
END A_H
PIN A_L
 # pin definition
END A_L
...
END AND2
```

```
MACRO AND2_5
FOREIGN AND2 0 0 ;
...
PIN Y_L
 # pin definition
END Y_L
PIN Y_H
 # pin definition
END Y_H
...
PIN A_L
 # pin definition
END A_L
PIN A_H
 # pin definition
END A_H
...
END AND2_5
```

Fig. 5.11 Abstract view and LEF macro description for AND2 master cell and its variant

- fringe capacitance, corresponding to the coupling between the vertical side of a wire and the neighboring routing layer

Each of these capacitances is defined per unit length of wire. Typically, capacitance values are extracted from detailed simulations (using a 3-D field solver), and tabulated as a function of wire width and wire spacing.

To the router, each differential pair will be represented as a single wire, with a width equal to the sum of both wire width and their spacing. When parasitics are extracted for such a logical wire, three corrections need to be made:

- the wire width is different from the physical width

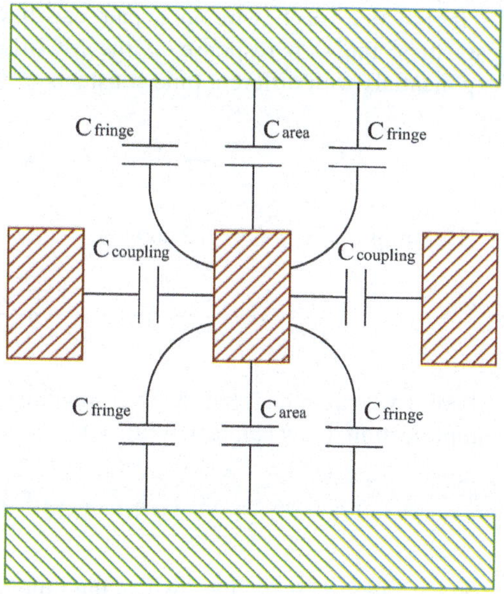

Fig. 5.12 Wire capacitance modeling for place and route

Table 5.2 Physical and logical capacitance parameters

Capacitance parameters		
Quantity	Physical	Logical
Area capacitance	$C_A(w)$	$C'_A(w') = C_A(w) + 2 \cdot C_C(w, P - W) + C_F(w, P - W)$
Coupling capacitance	$C_C(w, s)$	$C'_c(w', s) = \frac{1}{2} \cdot C_c(w, s)$
Fringe capacitance	$C_F(w, s)$	$C'_F(w', s) = \frac{1}{2} \cdot C_F(w, s)$

$w' = w + P$

- the coupling capacitance between the two physical wires of a same pair is not extracted
- the fringe capacitance due to internal sides of the differential pair wires is not extracted

In order to fix these issues, and taking into account the Miller effect in the coupling between wires of a same pair, the capacitances tables can be modified as shown in Table 5.2. In this transformation, a total capacitance is calculated as the sum of the capacitances of two wires in the pair—including coupling on both sides of the logical wire—and the total capacitance is divided by two to yield an average capacitance per wire. Therefore, the differential pair parasitics is modeled as the average of the parasitics of its two wires.

In practice, each wire in a pair can have a different amount of associated parasitics; the step response of the MCML circuit will then be a sum of two exponentials with different time constants

$$\frac{V_{o,d}}{2 \cdot V_{sw}} = 1 - \left[e^{-\frac{t}{\tau_1}} + e^{-\frac{t}{\tau_2}} \right] \tag{5.1}$$

Defining $\tau = (\tau_1 + \tau_2)/2$ and $\Delta\tau = (\tau_1 - \tau_2)/2$, we can write

$$\frac{t}{\tau_{1,2}} = \frac{t}{\tau \pm \Delta\tau} = t \cdot \frac{\tau \mp \Delta\tau}{\tau^2 - (\Delta\tau)^2} \approx \frac{t}{\tau} \cdot \left(1 \mp \frac{\Delta\tau}{\tau}\right) \tag{5.2}$$

where it was assumed that $\Delta\tau/\tau$ is small enough so that $(\Delta\tau/\tau)^2 \approx 0$. With this simplification, (5.1) can be rewritten as

$$\frac{V_{o,d}}{2 \cdot V_{sw}} \approx 1 - 2 \cdot e^{-\frac{t}{\tau}} \cdot \cosh\left(\frac{t}{\tau} \cdot \frac{\Delta\tau}{\tau}\right) \tag{5.3}$$

The cosh term is very close to 1 in the time range of interest, for small $\Delta\tau/\tau$, thus the delay is very close to $\tau \cdot \ln(2)$. No closed form expression can be written for the delay in the general case, but this can be verified from the numerical results plotted in Fig. 5.13.

As it can be seen, the delay variation due to unbalanced loading is less than 10% even for large $\Delta\tau/\tau$. Therefore, using the average time constant gives a very good approximation. In practice, the routing of true differential pairs as proposed above results in well-balanced time constants, and the error is limited to less than 1% in most of the cases, which is negligible compared to the accuracy of parasitics extraction anyway.

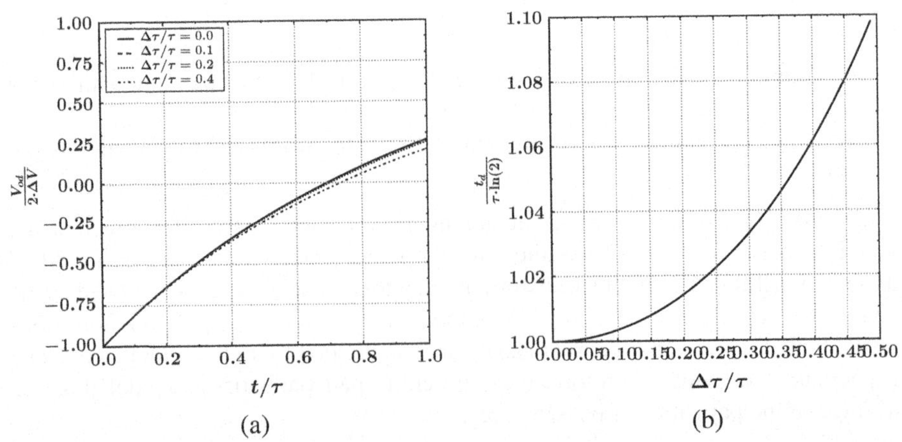

Fig. 5.13 Variation of the delay due to the unbalanced loading in an MCML gate (**a**) voltage waveforms according to Eq. (5.3) for different values of $\Delta\tau/\tau$ (**b**) delay variation as a function of $\Delta\tau/\tau$

Part III
Design Examples

Chapter 6
Design Example I: Low-Noise Encoder Circuit for A/D Converter

6.1 Circuit Description

The circuit to be implemented is a decoder block for an analog-to-digital converter. Without disclosing the details of the circuit functionality, it is sufficient to consider that it receives 23 bits from the core analog-to-digital part, and performs a complex arithmetical operation to calculate a 12 bits result. Additionally, several control inputs allow to switch the circuit into various operating modes.

The circuit has to operate at 1 GS/s, and to achieve this speed the original CMOS circuit implementation includes five parallel interleaved blocks as depicted in Fig. 6.1. A clock generator block generates five clock signals from the master 1 GHz clock, for each of the five decoder leaves, which trigger the processing of samples at 200 MS/s in a time-interleaved fashion.

6.2 MCML Cell Library

The MCML cell library implemented for this design and described in this section targets a 0.18 μm CMOS process, with 1.8 V supply voltage. For design-specific reasons and easier integration, the nominal supply voltage of 1.8 V has been used for the MCML cells. It should be noted that this does not necessarily give the best power-performance compromise. Moreover, the design complexity estimated at about 4k cells and strict area requirements dictated the use of a compact cell set (Table 6.1).

S. Badel et al., *Design Automation for Differential MOS Current-Mode Logic Circuits*, https://doi.org/10.1007/978-3-319-91307-0_6

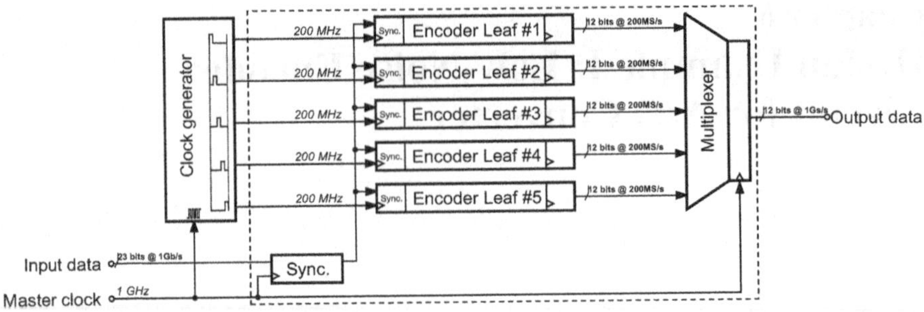

Fig. 6.1 5× interleaved encoder block diagram

Table 6.1 Summary of parameters for the MCML standard-cell library

Design parameters	
Supply voltage	1.8 V
Voltage swing	500 mV
Unit tail current	25 μA
Noise margin	100 mV
N bias voltage	Nom. 700 mV
P bias voltage	Nom. 150 mV
Layout parameters	
Cell height	14.4 μm
V_{DD}/GND rail width	1.84 μm
Load device size	0.68 × 0.82 μm
Current source size	0.68 × 0.9 μm
Routing pitch	0.96 μm

6.2.1 Library Parameters

The choice of the three main design parameters, namely the voltage swing, unit tail current, and noise margin, has a great impact on the resulting cell performance. These choices are detailed below.

Voltage Swing As it was pointed out several times, there are many benefits of increasing the voltage swing. MCML cells with higher voltage swing have lower intrinsic delays, exhibit smaller area, and produce less supply noise. Therefore, the voltage swing was chosen as large as practical, limited by the linearity of PMOS loads under worst-case process conditions.

Unit Tail Current Transistor sizes for the PMOS loads and NMOS current sources were chosen to be about three times larger than the minimum design rule, both in width and length, in order to provide sufficient matching. The exact sizes were adjusted for layout efficiency. According to these device sizes, the unit tail current was chosen as large as practical, limited by the maximum gate voltage of the NMOS current sources.

Table 6.2 Main functions in the standard-cell library and their drive strengths

Function	Drive strengths
BUF	1, 2, 3, 4, 5, 8, 11, 17, 32, 50, 64
AND2	1, 2, 4, 8, 12, 16
AND3	1, 2, 4, 8
AO21	1, 2, 4, 8
DFF	1, 2, 4, 8
MAJ32	1, 2, 4, 8
MUX2	1, 2, 4, 8, 12, 16
XOR2	1, 2, 4, 8, 12, 16
XOR3	1, 2, 4, 8

Noise Margin The noise margin was chosen to be 100 mV; according to the device sizes chosen earlier, and the matching characteristics of the process, this value guarantees a worst-case noise margin of about 50 mV.

6.2.2 Cell Selection

The library is composed primarily of basic functions: BUF, AND2, AND3, AO21, XOR2, XOR3, MAJ32, and MUX2, in addition to a master–slave latch. Seven different footprints were needed to implement this basic set of function. Based on the footprints, optimized layouts were created for the MAJ32, XOR2, and XOR3 functions. Each cell design branches out into 4 to 6 different drive strengths (11 for the buffer), resulting in 71 cells (Table 6.2).

6.2.3 Cell Characteristics

It is interesting to compare the characteristics of the MCML cells to their CMOS counterparts. Clearly, the different characteristics of both logic styles and the independent choice of parameters make it hard to establish a one-to-one comparison between cells in both libraries. In the plots presented hereafter, cells are compared with equivalent functionality and drive strength, but it must be highlighted that both criteria have different expressions in both libraries.

Area Figure 6.2 shows a scatter plot of the area of MCML cells versus the area of corresponding CMOS cells. It is not surprising that MCML cells are on average larger than their CMOS counterparts, due to the additional circuitry involved (load devices, current source). As this plot shows, the area of MCML cells is comprised between one and three times the area of CMOS cells.

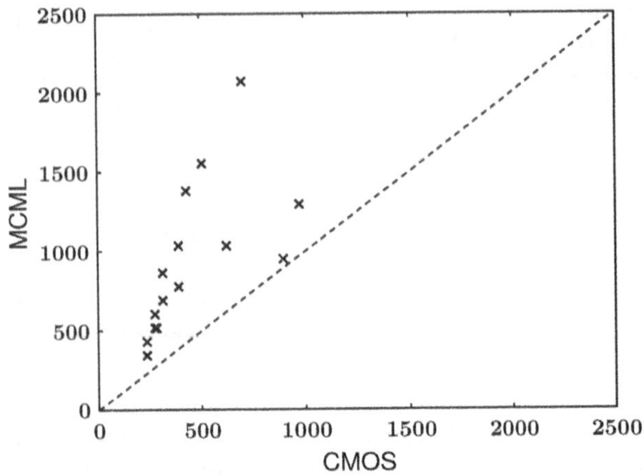

Fig. 6.2 Comparison of area of MCML and CMOS standard cells. Each point's x- and y-coordinates are the area (in μm^2) of an equivalent cell in the CMOS and MCML libraries, respectively

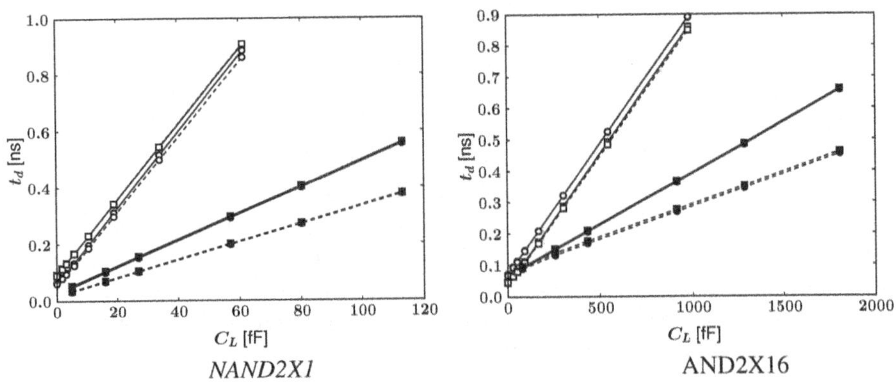

Fig. 6.3 Comparison of delay-load characteristics of MCML and CMOS standard cells. Black and white symbols denote CMOS and MCML cells, respectively; solid and dashed lines denote rising and falling transitions, respectively

Delay Figure 6.3 compares delay-load capacitance characteristics of equivalent MCML and CMOS cells. On each graph, the black symbols denote CMOS cells and the white symbols MCML cells, whereas the solid lines denote rising transition and the dashed lines denote falling transitions. As it can be seen, the drive strength of MCML gates is lower than the drive strength of their CMOS counterparts. The intrinsic delay, though, is in the same order of magnitude. A striking characteristics of MCML cells is the symmetry of their rise and fall times; even though the AND2 network is not fully symmetrical, the rise and fall times are very well balanced compared to CMOS cells.

Figure 6.4 compares the delay-input rise (fall) time characteristics of MCML and CMOS NAND2X1 gates. Here also, a sharp difference can be seen: in MCML gates, the delay depends very weakly on the slope of the input signal, compared to CMOS gates.

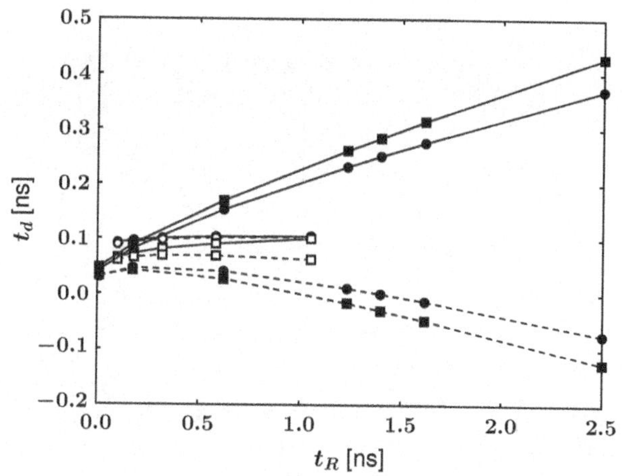

Fig. 6.4 Comparison of delay-input rise time characteristics of MCML and CMOS standard cells. Black and white symbols denote CMOS and MCML cells, respectively; solid and dashed lines denote rising and falling transitions, respectively

6.2.4 *Cell Layout*

Example layouts of AND2 cells for the two extreme drive strengths are plotted in Fig. 6.5. The PMOS load transistors are located at the top of the cells, near the power rail. Similarly, NMOS current sources are located near the ground rail, at the bottom of the cell. Above the power rail and below the ground rail are the bias rails; these rails are of near-minimal size and are shared by two adjacent rows when the cells are flipped and abutted.

6.2.5 *Bias Generator*

The bias voltages are generated by two replica generators, based on one external reference voltage and one external reference current (Fig. 6.6).

6.2.6 *Level Converters*

The circuit requires level converter to interface it with existing blocks using CMOS levels, and to save I/O pads for driving the output signals off-chip.

The CMOS-to-MCML converter (Fig. 6.7a) is implemented by a DCVSL buffer providing complementary input signals to an MCML buffer. The DCVSL gate allows to keep the common-mode voltage above $V_{DD}/2$, in order to avoid pulling the tail current source of the MCML buffer out of saturation during switching.

The MCML-to-CMOS conversion is achieved by an OTA stage followed by a regular CMOS inverter that restores the rail-to-rail logic levels (Fig. 6.7b).

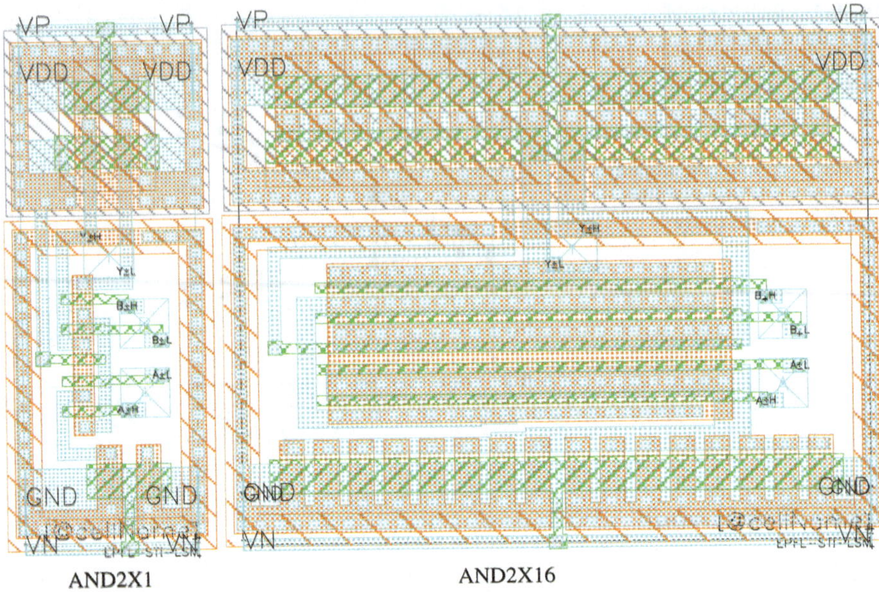

Fig. 6.5 Example layouts of AND2 standard cells, for drive strengths ×1 and ×16

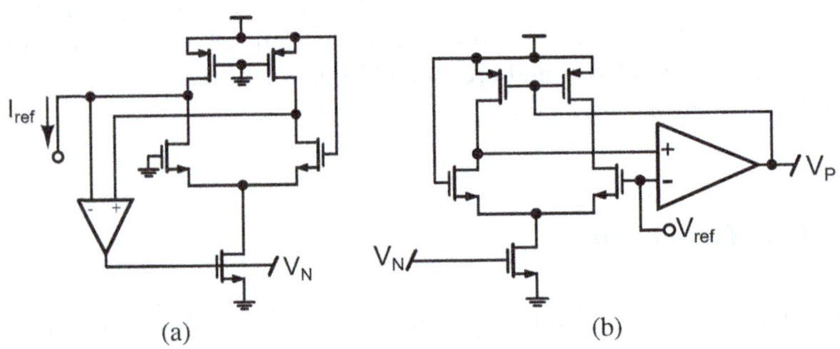

Fig. 6.6 Implemented bias generator (**a**) replica bias for tail current (**b**) replica bias for voltage swing

6.3 Design Flow

A variety of tools needed to be created to implement the design flow are described in Chap. 5. The front-end part of the flow, illustrated in Fig. 6.8, includes the creation of the standard cells, and the generation of various data for the back-end tools. In this part, the custom tools (indicated with shaded rectangles on the figure) were created to perform the cell characterization (timing and power data extraction), and to generate the logical views (indicated with dotted borders on the figure) for the various types of data.

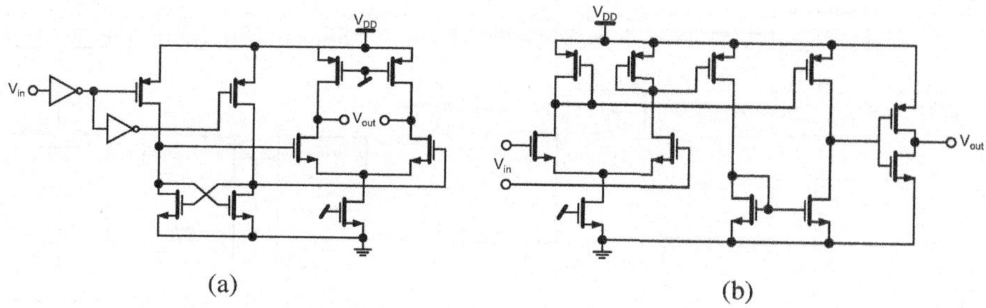

Fig. 6.7 Implemented level converters (**a**) CMOS-to-MCML converter (**b**) MCML-to-CMOS converter

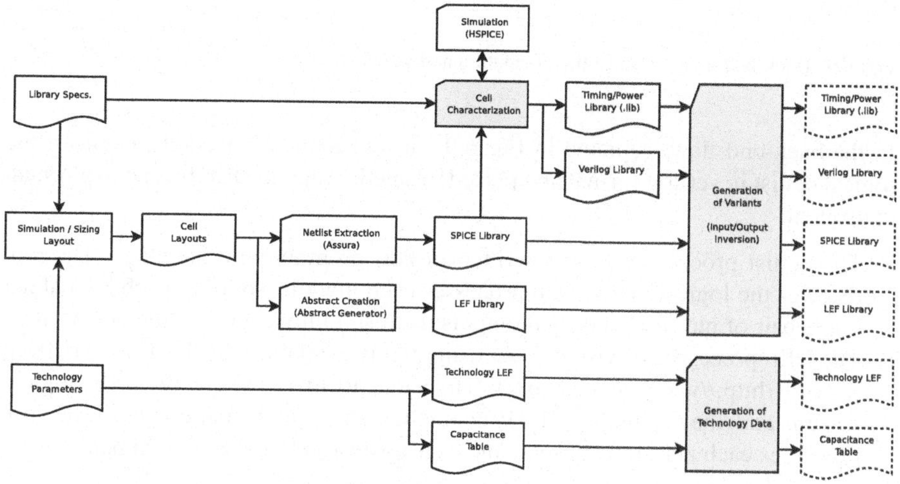

Fig. 6.8 Front-end flow: cell library creation and data generation

Though it is composed of three distinct parts on the figure, these three parts are integrated into a common tool written in Perl programming language. In this way, the consistency is guaranteed amongst the many different output files.

- The input to this tool are the SPICE netlists of the cells, along with a description of their functionality. The tool generates test benches for cell characterization, runs the simulator, and collects the results in a database. Based on this data, the tool can output verilog models of the cell library, and Liberty library for synthesis and P&R.
- Additionally, the tool can generate logical libraries, including variants of the cells with inverted input/output pins, in Verilog, SPICE, Liberty and LEF formats based on the physical data and the cell descriptions.
- For technology data generation, the tool processes the technology LEF and capacitance table according to user-defined parameters for differential routing and generates the corresponding files with modified parameters.

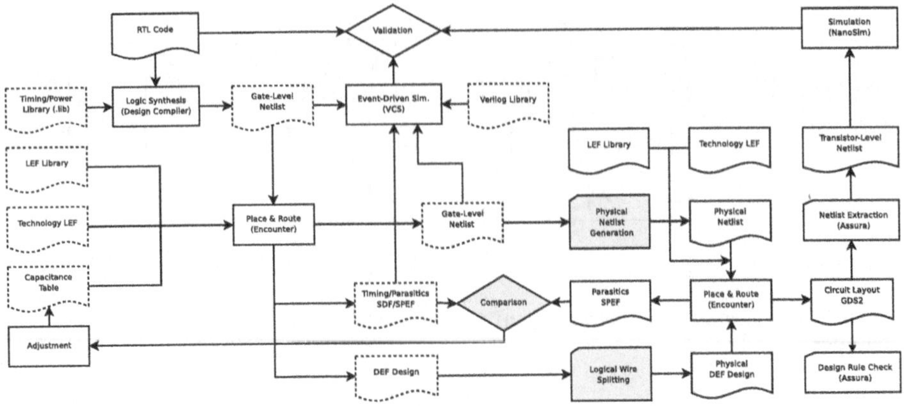

Fig. 6.9 Back-end flow: circuit implementation and verification

In the back-end flow, depicted in Fig. 6.9, custom tools are needed to convert the logical netlist in verilog format and placed&routed design in DEF format to physical equivalent.

- The netlist processing tool is written in Perl programming language. The tool receives the logical netlist, and processes each module, splitting each logical net into a pair of physical nets, and outputs the result into a new verilog netlist file.
- The DEF processor is written in C, using freely available LEF/DEF parsers from Si2.org (http://www.openeda.org). The input to this tool are the DEF logical design, and the technology LEF containing the routing parameters. The tool processes each net in the design, and transforms it into a differential pair.

Starting from the RTL code, the circuit is synthesized using the logical library data. The synthesized netlist is then placed and routed, using the logical technology data. The resulting netlist and DEF design are converted to physical equivalents, and fed back to the router using the physical technology and library data, to complete the missing connections.

Additionally, a third back-end tool was created for comparing sign-off parasitics to the parasitics extracted from the logical layout. This tool processes the parasitics file in SPEF format extracted from the physical circuit layout, identifies differential pairs, and calculates the average capacitive load for each pair including Miller effect. Then, this data is compared to the parasitics extracted from the logical circuit layout, allowing to adjust the capacitance tables for better correlation with the sign-off results.

6.4 Results

6.4.1 Encoder Redesign

The described circuit was implemented using the standard-cell library, through the proposed design flow. The resulting circuit layout is shown in Fig. 6.10. Figure 6.11 features a close-up view of the layout, showing the routing.

Post-layout simulations have been performed on the implemented design, confirming its correct functionality. The area, power consumption, and power supply noise of the MCML circuit are compared with the original CMOS implementation in Table 6.3.

As the results display, the MCML encoder consumes about 9.7 times more power than the CMOS encoder, due to the static power dissipated in MCML gates. In terms of area, the MCML circuit is larger by a factor of 2.2, due to the larger sizes of the gates. Note that the area is not fully utilized in the MCML block, since its size and the pin positions were fixed in advance, and decrease in area could thus be achieved. The peak–peak supply noise, obtained by post-layout simulations, is about 30 times smaller in the MCML circuit than in the CMOS circuit, despite its much higher power consumption.

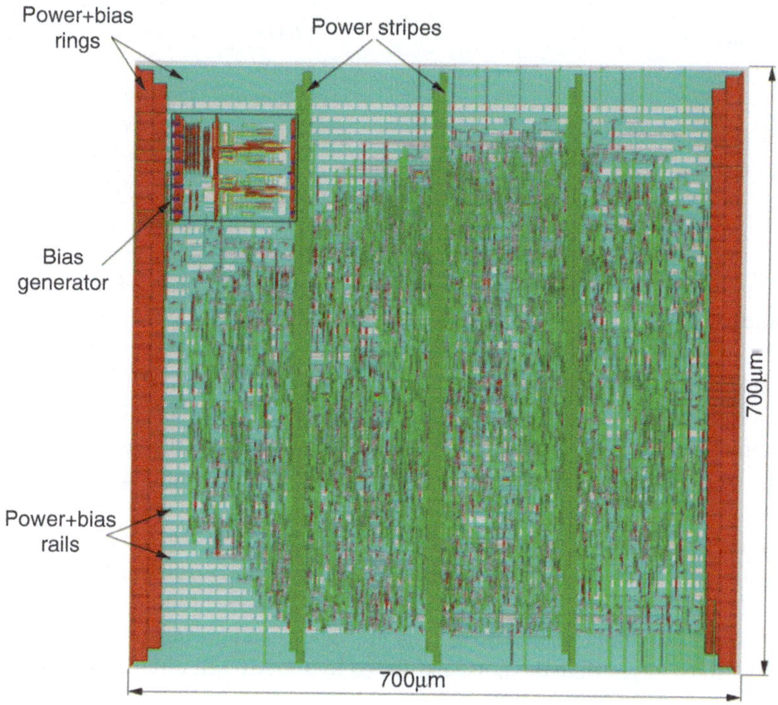

Fig. 6.10 Layout view of the implemented MCML encoder

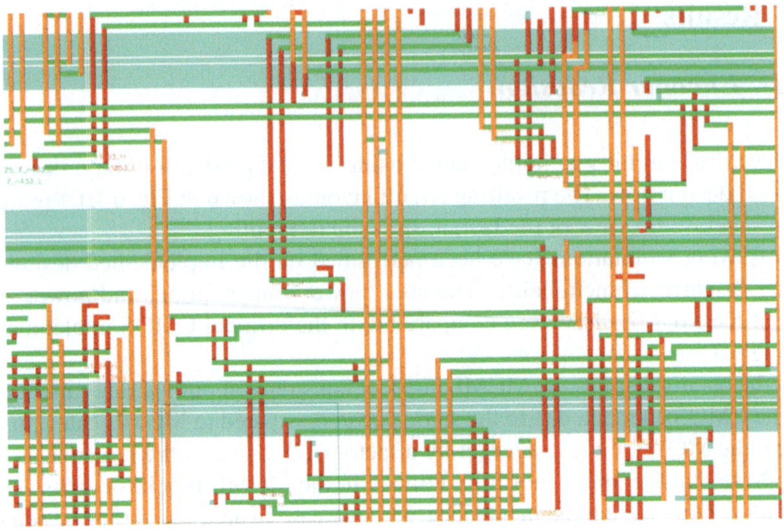

Fig. 6.11 Close-up view of the MCML encoder layout showing differential routes

Table 6.3 Implementation results of MCML encoder compared to the original CMOS implementation

	CMOS circuit	MCML circuit
Clock frequency	1 GHz	1 GHz
Power consumption	30 mW	290 mW
Area	0.22 mm^2	0.49 mm^2
Peak–peak AC power supply current	~10 mA	~300 μA
Peak–peak V_{DD} voltage noise with 1 nH inductance	~80 mV	~650 μV
Peak–peak GND voltage noise with 1 nH inductance	~60 mV	~600 μV

The gains in noise performance are illustrated in Figs. 6.12 and 6.13. Figure 6.12 plots the total AC supply current of the MCML and CMOS encoder blocks, obtained by post-layout simulations. Figure 6.13 shows the frequency spectrum of the supply noise power, where it can be seen that the magnitude of the first harmonic of the clock frequency is reduced by about three orders of magnitude.

In practice, the varying current on the power supply creates a noise voltage through the resistance of the power networks and the inductance of package bondwires. Figures 6.14 and 6.15 show the respective voltage at the V_{DD} and GND pins when the circuit is simulated with series resistance of 1 nH, a typical value for bondwire inductance. In these simulations, only the clock is running while the inputs are stable, minimizing the switching activity. The CMOS circuit exhibits peak–peak noise voltages of about 80 and 60 mV on the supply and ground pins, which is more than 100 times larger than the noise exhibited by the MCML circuit, with peak–peak

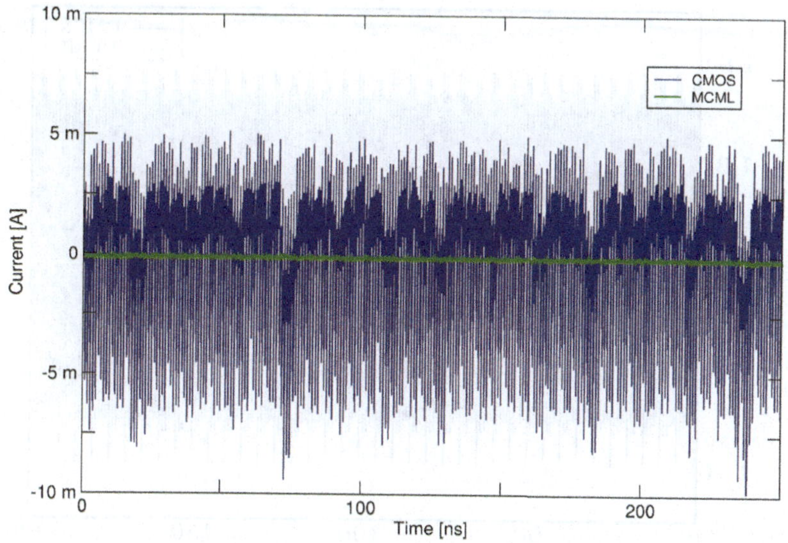

Fig. 6.12 Post-layout simulated AC power supply current noise of the MCML and CMOS encoders, showing a reduction of noise power by about three orders of magnitude at the operating frequency

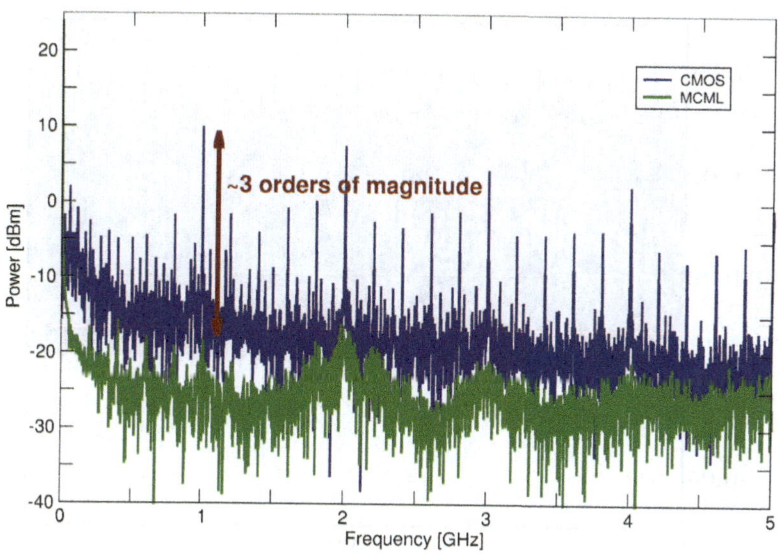

Fig. 6.13 Post-layout SPICE simulated AC power supply noise spectrum of the MCML and CMOS encoders

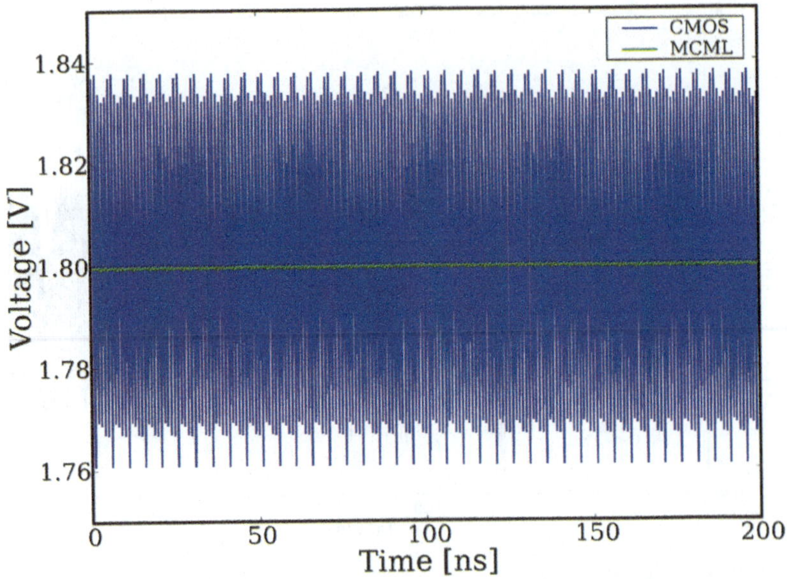

Fig. 6.14 SPICE simulated voltage noise on the power supply of the MCML and CMOS encoders with a 1 nH bondwire inductance

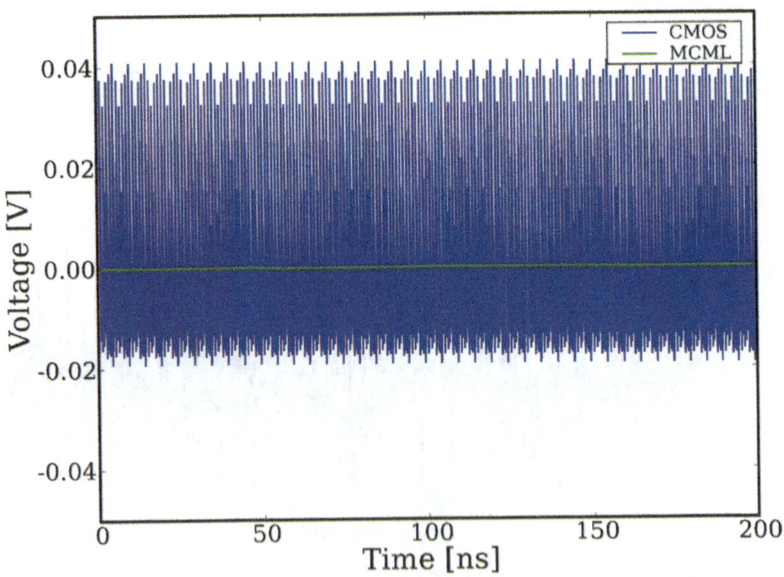

Fig. 6.15 SPICE simulated voltage noise on the ground supply of the MCML and CMOS encoders with a 1 nH bondwire inductance

values of 650 and 600 μV, respectively. This highlights the benefits of the MCML circuit style, as well as the role of the clock network in the generation of digital switching noise.

6.4.2 Architecture Modification

The encoder presented in the previous section was redesigned with the same architecture as the original CMOS encoder. In CMOS circuits, since the power is proportional to the switching parallel implementations of circuits running at lower speed do not dissipate more power, except for the additional circuitry needed to interleave the parallel circuits. However, in MCML circuits, the power consumption does not depend on the switching frequency, and to achieve better power efficiency, gates must be run at the highest possible activity factor.

In practice, this means that pipelined circuits with short pipeline stages will be more efficient than parallel implementations. To verify this, the encoder circuit RTL code was modified to parameterize the number of parallel leaves, and the design was implemented with different number of leaves and different number of pipeline stages per leaf. The results are shown in Table 6.4. As it can be seen from these results, the designs using a single leaf and a larger number of pipeline stages have better area and power performance. In fact, the circuit area can be reduced close to that of the CMOS implementation through pipelining. The lowest power is achieved with one parallel leaves and four pipeline stages. Note that the power dissipation of the MCML implementation can be reduced by a factor of up to 2.6 through pipelining, without influencing the superior noise performance. In all cases however, the power consumption is still more than three times larger than the power dissipated by the CMOS block.

Table 6.4 Implementation results of MCML encoder performance (area and power) for different architectures that exploit pipelining

Parallel leaves	Pipeline stages	Area[a] (mm)	Power (mW)
1	4	0.17	110
1	5	0.20	130
2	2	0.23	155
2	3	0.21	135
3	1	0.37	380
3	2	0.28	160
4	1	0.34	285
4	2	0.31	180
5	1	0.30	290

[a]Useful area = total area × cell density

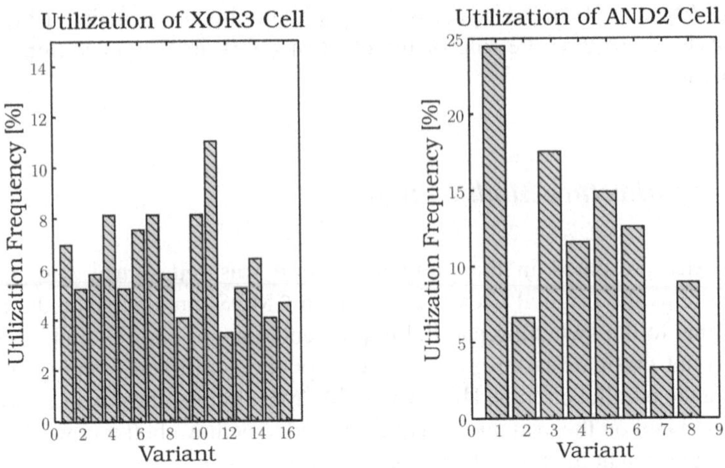

Fig. 6.16 Utilization of the cell variants with inverted inputs/output of the XOR3 and AND2 cells

6.4.3 *Design Flow*

Some statistics have been extracted from the implemented design in order to evaluate the efficiency of the design flow. The results are presented in the following paragraphs.

6.4.3.1 Synthesis

Figure 6.16 illustrates the utilization by the synthesizer of the different variants with inverted inputs/output for the XOR3 and AND2 cells. The statistics display a balanced utilization of all the different variants, indicating a very good utilization of the free inversions by the tool.

6.4.3.2 Differential Routing

In order to evaluate the efficiency of the differential routing, the resulting design has been compared against the same design implemented with a regular flow, not enforcing the routing of differential wire pairs. The plots in Fig. 6.17 display the extracted relative parasitic capacitance mismatch between the two wires in a differential pair for both approaches. As it can be seen, the differential approach results in better matched parasitics, with more than 60% of the nets exhibiting a relative mismatch of less than 10% and about 90% of the nets with parasitics matched to 10%, which is beneficial both to the timing integrity and the noise performance. In contrast, the design routed with a conventional flow displays more spread in the distribution, with a significant amount of nets with mismatch larger

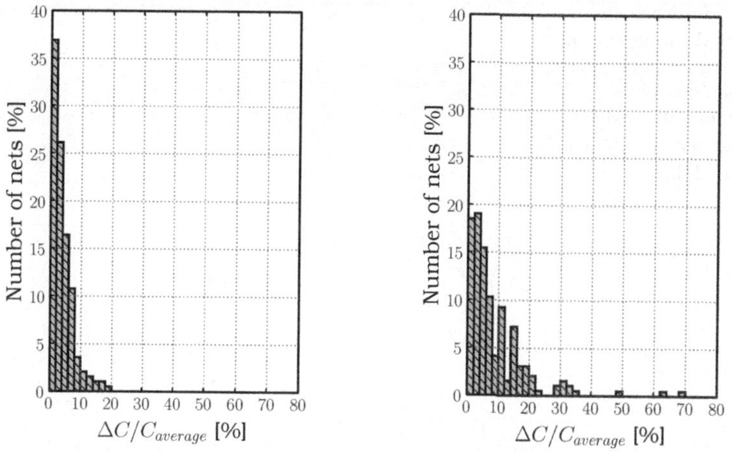

Fig. 6.17 Statistical distribution of the mismatch in extracted parasitic capacitance between two wires of a differential pair. Left: design routed with the proposed flow; right: design routed with a conventional single-ended flow

than 20% and occasionally up to 70%. The very limited mismatch between the differential wires also results in balanced switching characteristics, and thus reduced switching noise, which contrasts with the results obtained with conventional routing.

The differential routing efficiency is also illustrated in Fig. 6.18, where the routing of a portion of a clock net is extracted from the same two designs. In the design routed with a conventional approach, not enforcing the parallel routing of differential pairs, the resulting routing displays a large average separation between the two wires. This results in unbalanced loading, which adversely affects the switching noise performance. Additionally, this can cause skew between the two signal polarities, which can cause the differential clock signal to stay at an undefined logic level for a period of time, resulting in an increase of the setup and hold times. Finally, this results in a significant increase of the self-inductance due to the large loop area enclosed by the wire pair [1], which increases the magnetic coupling to other wires and decreases the crosstalk immunity.

6.4.3.3 Parasitics Modeling

The accuracy of the parasitics estimation is reported in Fig. 6.19, which plots the estimated parasitic capacitance extracted natively from the P&R tool using the model described in Chap. 5 against the parasitics extracted from the final layout using an accurate sign-off parasitics extraction tool. Sign-off parasitics have been processed by locating the two components of each differential pair and calculating the average capacitance, with a multiplying factor of two for the coupling capacitance between the two components modeling the Miller effect. It can be seen that the extracted values match the modeled values very accurately, in a vast majority of cases within ±15% of the expected value.

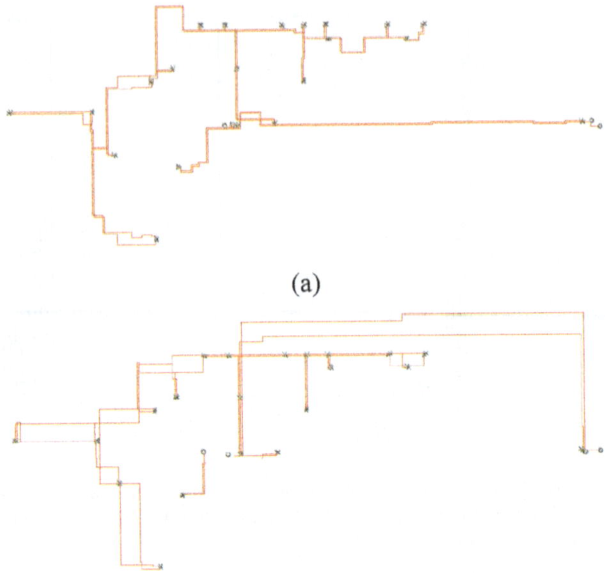

(a)

(b)

Fig. 6.18 Screenshots illustrating the routing of part of a differential clock net (**a**) design routed with the proposed flow (**b**) design routed with a conventional single-ended flow

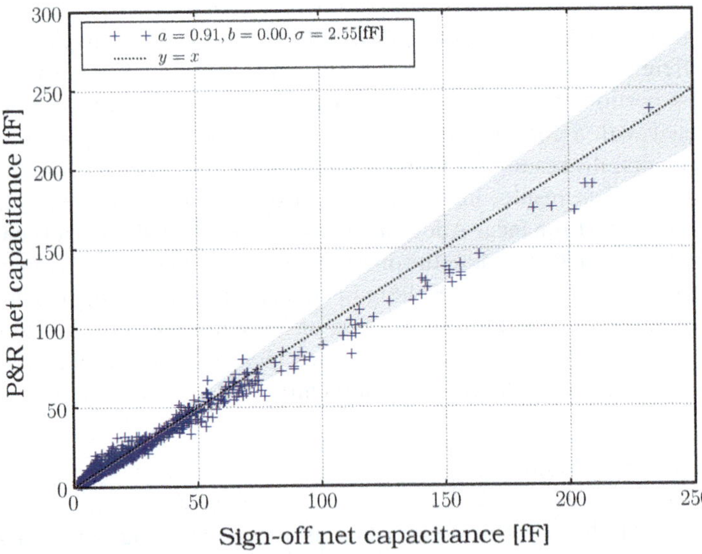

Fig. 6.19 Correlation between the P&R and sign-off extracted parasitic capacitances. The shaded area represents the ±15% accuracy interval

Reference

1. İ. Hatırnaz, A new interconnect-centric design methodology for high-speed standard cells with crosstalk immunity. PhD thesis, Ecole Polytechnique Fédérale de Lausanne, 2006

Chapter 7
Design Example II: High-Speed Multiplexer

7.1 Circuit Description

Due to their lower intrinsic delay when compared to CMOS circuits, MCML circuit offer fast operating speed when operated at low C_L/I_{SS} ratio. In order to demonstrate this, a relatively simple circuit is implemented. The proposed circuit is a pipelined 2^N-to-1 multiplexer, described by the RTL code featured on Listing 7.1 on page 152. The equivalent circuit schematic, is pictured in Fig. 7.1, consists of a multiplexer tree with pipeline stages at each level of the tree. Since this circuit has extremely short pipeline stages, it can run at very high speed. In addition, short interconnections result in a significant part of the logic delay being due to the gate internal delays. The speed of such a circuit will therefore be limited mostly by gate delays and register setup times, as well as clock skew.

Such a simple circuit could clearly be handcrafted for maximum performance. The use of an automated approach is nevertheless interesting in such a case, in order to benefit from numerous advantages provided by a semi-custom design flow, namely:

- the use of digital-specific tools such as static timing analysis, or clock tree synthesis
- the easy and fast turnaround when design changes are required
- the seamless integration within a digital top-down design flow

7.2 MCML Cell Library

The minimal MCML cell library implemented for this design is optimized for high-speed operation. The target technology is a 90 nm CMOS process with a metal stack of nine layers, including six fine-pitch routing layers.

S. Badel et al., *Design Automation for Differential MOS Current-Mode Logic Circuits*, https://doi.org/10.1007/978-3-319-91307-0_7

```
//------------------------------------------------------------------------------
// Module : multiplexer
//------------------------------------------------------------------------------

module multiplexer
/** Parameters **/
#(
 parameter N = 4  // log2 of the number of input channels
)
/** I/O Ports **/
(
 // Inputs
 input          clk,      // input clock
 input [2**N-1:0]  d_in,  // input data (2**N bits)
 input [N-1:0]     sel,   // select signal
 // Outputs
 output reg    d_out      // output data (1 bit)
);

// Signals
reg [2**N-1:0]  d_in_t;
reg [2**(N-1)-1:0] d_p [N-2:0]; // data pipeline registers
reg [N-1:0] sel_t;              // timed select registers

// Input data sampling
always @( posedge clk ) begin
 d_in_t <= d_in;
end

// Multiplexer tree
generate
genvar k, l;
 for( k = 0 ; k <= N-1 ; k=k+1 )
 begin: mux_stage
  // select pipeline
  if ( k > 0 ) begin
   reg [k-1:0] sel_p;
   for( l=0; l<=k; l=l+1)
   begin: sel_unit
    if( l == 0 ) begin // first stage
     always @( posedge clk )
      sel_p[l] <= sel[k];
    end else if( l < k ) begin // middle stage(s)
     always @( posedge clk )
      sel_p[l] <= sel_p[l-1];
    end else begin // last stage
     always @( posedge clk )
      sel_t[k] <= sel_p[l-1];
    end
   end
  end else begin /* k == 0 */
   always @( posedge clk )
    sel_t[k] <= sel[k];
  end

  // multiplexers
  for( l=0; l<2**((N-1)-k) ; l=l+1 )
  begin: mux_unit
   if( k == 0 ) begin // first stage
    always @( posedge clk )
     d_p[k][l] <= sel_t[k] ? d_in_t[2*l+1] : d_in_t[2*l];
   end else if( k < N-1 ) begin // middle stage(s)
    always @( posedge clk )
     d_p[k][l] <= sel_t[k] ? d_p[k-1][2*l+1] : d_p[k-1][2*l];
   end else begin // last stage
    always @( posedge clk )
     d_out <= sel_t[k] ? d_p[k-1][2*l+1] : d_p[k-1][2*l];
   end
  end
 end
endgenerate

endmodule /* multiplexer */
```

Listing 7.1 RTL description of the 2^N-to-1 multiplexer

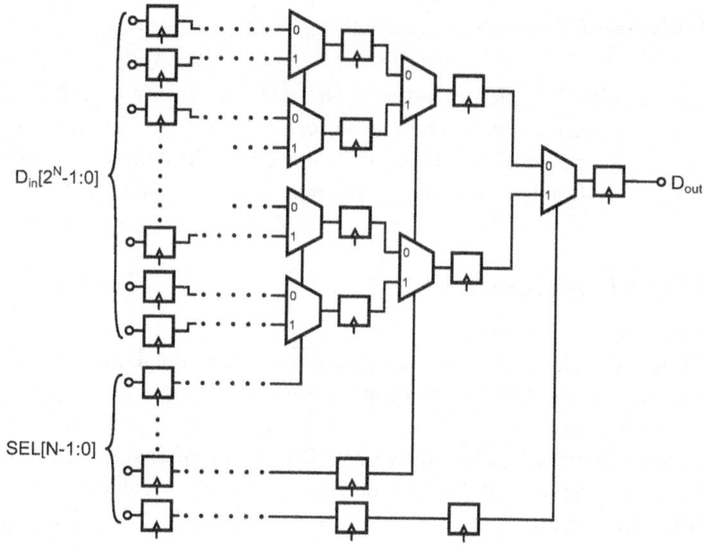

Fig. 7.1 Equivalent circuit schematic for the 2^N-to-1 multiplexer

Table 7.1 Summary of parameters for the MCML standard-cell library

Design parameters	
Supply voltage	1.0 V
Voltage swing	400 mV
Unit tail current	250 μA
Noise margin	>20 mV
N bias voltage	Nom. 650 mV
P bias voltage	Nom. 240 mV
Layout parameters	
Cell height	13.44 μm
V_{DD}/GND rail width	2.39 μm
Load device size	1.8 × 0.08 μm
Current source size	4.1 × 0.125 μm
Routing pitch	0.28/0.42 μm

7.2.1 Library Parameters

The process nominal supply voltage of 1 V has been used for the MCML cells, resulting is small size for the PMOS load device and high operating speed. The voltage swing is chosen to be as large as practical, in order to provide the highest operating speed. In order to operate at low C_L/I_{SS} ratio, the unit tail current is chosen as large as 250 μA. Based on these parameters, transistor sizes are adjusted through Monte-Carlo simulation in order to provide a worst-case noise margin of about 20 mV. The library parameters are summarized in Table 7.1.

7.2.2 *Cell Selection*

Due to the predictable implementation of the target circuit, a minimal set of cells compose the cell library. These include buffers in three drive strength, mainly for clock distribution, master–slave latch, and multiplexer in two drive strength. Two-input AND and XOR gates were also implemented for completeness (Table 7.2).

7.2.3 *Cell Characteristics*

The resulting cell characteristics are compared with the equivalent cells from a CMOS standard-cell library. Aiming at high speed operation, we are mostly interested in the intrinsic gate delays, and setup time of the flip-flops. The respective values are given in Table 7.3 for equivalent CMOS and MCML gates. Note that the values given are indicative of the type of gate, while the actual value can vary with the particular drive strength.

Table 7.2 Main functions in the standard-cell library and their drive strengths

Function	Drive strengths
BUF	1, 2, 4
AND2	1, 2
DFF	1, 2
XOR2	1, 2

Table 7.3 Comparison of MCML and CMOS gate intrinsic delay

Gate	Arc	CMOS (ps)	MCML (ps)
INV	I → O	5.6–15.8	5.6
BUF	I → O	19.8–50.7	5.6
NAND2	I1 → O	7.1–19.7	9.2
	I2 → O	9.6–26.5	20.8
AND2	I1 → O	21.5–56.8	9.5
	I2 → O	24.8–65.4	20.8
NOR2	I1 → O	10.8–30.0	9.5
	I2 → O	14.1–41.6	20.8
OR2	I1 → O	33.8–88.0	9.5
	I2 → O	36.1–96.6	20.8
XNOR2	I1 → O	16.7–44.2	5.8
	I2 → O	22.9–65.7	24.7
XOR2	I1 → O	16.9–44.7	5.8
	I2 → O	21.0–62.1	24.7
MUX2	D1 → O	44.2–122.6	12.5
	D2 → O	42.7–122.8	12.5
	SEL → O	44.8–139.4	23.7
DFF	CK → Q	64.7–175.5	31.4
	Setup time	49.1–153.9	27.2

Values for CMOS are given as minimum–maximum couples, based on the corner cases for the process. The typical value is given for MCML gates, since the variation of the delay due to process variations is negligible. It should be noted that the delay of MCML gate can increase due to on-chip variations, and the variability of the delay will be larger for smaller drive strengths. Monte-Carlo simulation with smaller drive strength gates gives a relative standard deviation of about 5.2%, corresponding to a worst-case (3σ) delay about 16% higher than the nominal delay.

From the number given in Table 7.3, the speed advantage of MCML compared to CMOS is clear. Gate delays are between one and five times lower for MCML gates than for their CMOS counterparts, even when some margin is added to account for the worst-case mismatch. This can be achieved by the use of a large unit bias current, thus relatively large unit gate size, allowing to minimize the influence of intra-cell wiring and diffusion parasitics. In addition, the use of low-V_T transistors for the PMOS load devices allows to reduce their size and associated drain parasitics. Finally, note that the CMOS gates which are non-inverting incur a delay overhead due to the need to integrate an inverter stage in the gate, which is not necessary with MCML gates.

7.3 Implementation Results

Figure 7.2 displays the mask layout of the implemented MCML 16-to-1 multiplexer. The circuit was synthesized from the RTL description, then placed and routed according to the design flow described in the previous sections. A clock tree has been synthesized so as to provide a low skew clock to the pipeline register and

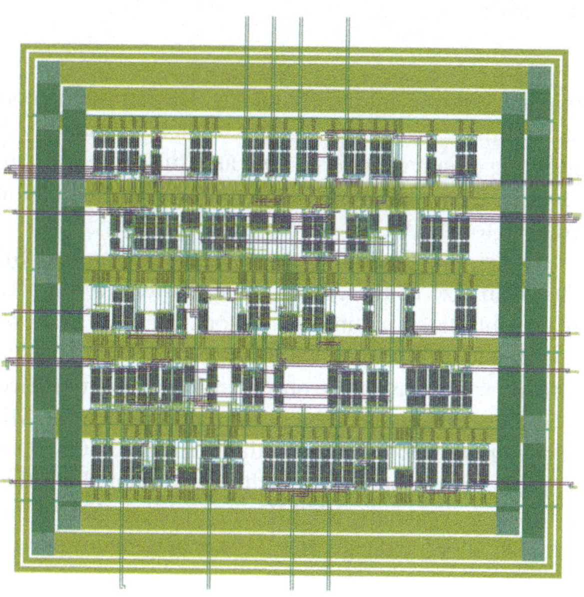

Fig. 7.2 Layout view of the implemented MCML multiplexer

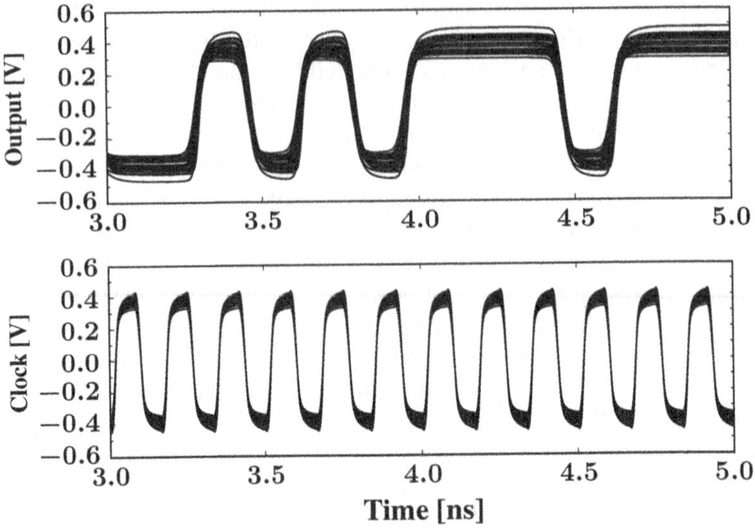

Fig. 7.3 Post-layout SPICE-simulated functionality of the MCML encoder circuit. Results of 50 Monte-Carlo iterations (with random process variations and mismatch) are superimposed

Table 7.4 Implementation results of MCML multiplexer compared to the CMOS implementation

	CMOS circuit	MCML circuit
Supply voltage (V)	1.2	1.0
Max. clock frequency (GHz)	3	6
Power consumption (mW)	8.3	24
Area (mm^2)	0.05	0.1
Cell count	106	76

maximize the operating frequency. During the complete flow, derating is used for static timing analysis to account for delay variability due to on-chip variations. The resulting circuit occupies a core area of $100 \times 100\,\mu m^2$ and operates at a maximum clock frequency of 6 GHz. The static power dissipation of the 76 gates in the circuit adds up to 24 mW.

The circuit functionality was verified through SPICE simulations. Thanks to the small size of the circuit, it is possible to run Monte-Carlo simulations with a reasonably large number of vectors. Figure 7.3 shows the simulated output of the circuit under 50 Monte-Carlo iterations. The variation in the voltage swing due to on-chip variations can be observed, but it does not affect the operation of the circuits.

Table 7.4 compares the performance of the MCML multiplexer with that of the same circuit implemented with a regular CMOS standard-cell library. The CMOS implementation can only reach a maximum clock frequency of 3 GHz, that is, two times slower than the MCML circuit (even though the CMOS circuit is operated at a higher supply voltage of 1.2 V for maximum speed).

Chapter 8
Design Example III: Grain-128a Stream Cipher

8.1 MCML for Cryptographic Applications

Due to higher and cheaper integration capabilities in semiconductor technologies, most of the daily life needs started to be solved by electronic systems. The sharp decrease of the cost of these microelectronic devices made the manufacturers produce quite attractive products for a large scale of customer segment and for different markets, like consumer electronics, military, space, etc. This abrupt progression in the microelectronic world also brought some problems with it. Since many of the electronic devices (such as mobile phones, RFID tags, wireless payment systems, etc.) have also the property of communication through common insecure mediums, the clients started to share many of their personal data after being processed with some complex security algorithms (encryption/decryption) with other users by using individual secret keys.

Although the security algorithms are quite secure in the first release and have strong mathematical proofs, many of them started to be mathematically broken after some time. For the ones which cannot be mathematically broken for the time being, physical hacking techniques (such as differential power analysis (DPA), electromagnetic attack, laser attack, reverse engineering, etc.) are discovered and used to extract the secure key which is stored inside the chip.

These hacking techniques (especially DPA and EM attacks) showed quite fast and successful results for the cryptographic circuits which are implemented with CMOS logic gates. Observation of the current consumption and radiation of the chips led these techniques show positive results, mainly because of high dynamic power consumption and high electromagnetic radiation properties of the single ended CMOS logic.

Due to the fact that CMOS logic circuits cannot hide the signature during the data process, new techniques and solutions were started to be searched for designing more robust cryptographic circuits. The fact that the digital circuits which are implemented with MCML gates have ideally no dynamic current consumption

S. Badel et al., *Design Automation for Differential MOS Current-Mode Logic Circuits*, https://doi.org/10.1007/978-3-319-91307-0_8

makes them interesting for security applications. Another interesting aspect of MCML is that the radiation due to data transitions is much less due to the closely coupled differential routing.

Considering the advantages explained above, a well-known cryptographic algorithm called Grain-128a is implemented using the MCML standard cells to see the differences and advantages on the CMOS implementation in terms of robustness under attacks. During the implementation, it is also targeted to characterize the switching noise performance of MCML gates by exploring their analog behaviors.

8.2 Circuit Description

In the digital world, **cipher** is an algorithm that takes a number of data bits (plain text), performs encryption and decryption to prevent them to be understood by anyone. The encryption and decryption are performed with a key which consists of M bits. There are two different types of ciphers [3] which are *Block Ciphers* and *Stream Ciphers*:

(a) **Block Ciphers:** A fixed number (N) of plain text bits are processed (encrypted/decrypted) in one cycle of the algorithm in a parallel manner (Fig. 8.1a). One cycle is usually more than one clock period.
(b) **Stream Ciphers:** The data bits are serially processed (encrypted/decrypted) in one cycle of the algorithm (Fig. 8.1b). One cycle is usually one clock period.

Grain-128a [1] is actually a new version of Grain-128 [2] with some plug-ins. The main purpose of Grain-128 and Grain-128a is to generate a bitstream (which is called keystream) which is pseudo-random. The main components of Grain-128 are two 128-bit shift registers (Fig. 8.2). The inputs of the shift registers are determined by some feedback functions. These feedback functions are combinational Boolean logic functions and their inputs are some specific bits of the shift registers. The names of the shift registers depend on the fact that whether their feedback functions are linear or nonlinear. Linear feedback functions are completely composed of XOR gates. If this is not the case, the feedback function is nonlinear. Therefore, the registers are called linear FSR (LFSR) and nonlinear FSR (NFSR). The contents of the LFSR are denoted by s_i, $s_{i+1}, \ldots, s_{i+127}$ where the subscript i denotes the quantization time. In the same manner the contents of the NFSR are denoted by b_i, $b_{i+1}, \ldots, s_{i+127}$.

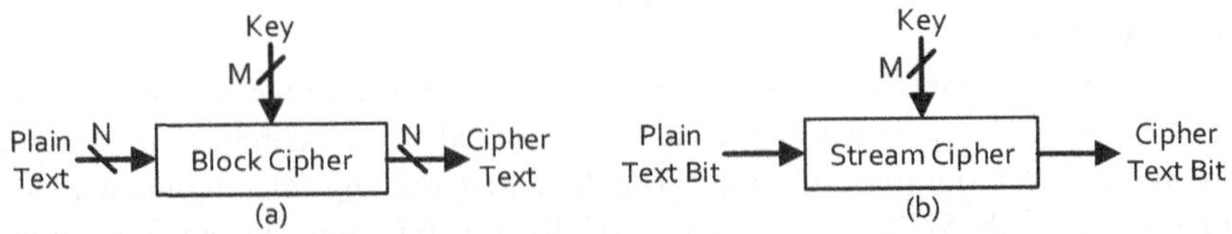

Fig. 8.1 (**a**) Block cipher, (**b**) stream cipher

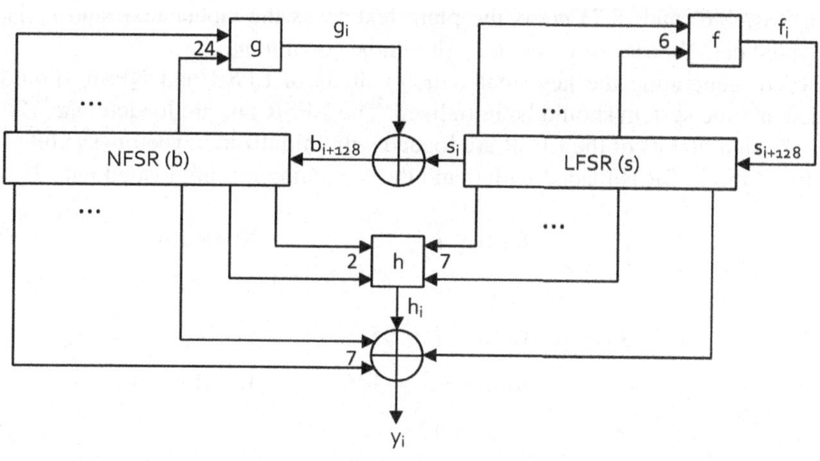

Fig. 8.2 Block diagram of Grain-128a algorithm

The next input of the LFSR is determined by the f function with the relation below:

$$f_i = s_{i+128} = s_i + s_{i+7} + s_{i+38} + s_{i+70} + s_{i+81} + s_{i+96} \tag{8.1}$$

The next input of the NFSR is determined by the g function with an additional XOR operation with the relations below:

$$\begin{aligned} g_i = {} & b_i + b_{i+26} + b_{i+56} + b_{i+91} + b_{i+96} + b_{i+3}b_{i+67} + b_{i+11}b_{i+13} \\ & + b_{i+17}b_{i+18} + b_{i+27}b_{i+59} + b_{i+40}b_{i+48} + b_{i+61}b_{i+65} + b_{i+68}b_{i+84} \\ & + b_{i+22}b_{i+24}b_{i+25} + b_{i+70}b_{i+78}b_{i+82} + b_{i+88}b_{i+92}b_{i+93}b_{i+95} \end{aligned} \tag{8.2}$$

$$b_{i+128} = s_i + g_i \tag{8.3}$$

Finally, the keystream which is denoted by y_i is determined by the h function and some more XOR operations:

$$h_i = b_{i+12}s_{i+8} + s_{i+13}s_{i+20} + b_{i+95}s_{i+42} + s_{i+60}s_{i+79}s_{i+94} \tag{8.4}$$

$$y_i = h_i + s_{i+93} + b_{i+2} + b_{i+15} + b_{i+36} + b_{i+45} + b_{i+64} + b_{i+73} + b_{i+89} \tag{8.5}$$

The encryption and decryption are performed simply by applying the plain text and the keystream to a 2-input XOR logic gate.

$$e_i = m_i + y_i \qquad \text{Encryption} \tag{8.6}$$

$$d_i = e_i + y_i \qquad \text{Decryption} \tag{8.7}$$

In Eqs. (8.6) and (8.7) m_i is the plain text, e_i is the cipher text, and d_i is the decrypted data. In case of no error d_i should be equal to m_i.

Before generating the keystream, the contents of LFSR and NFSR should be loaded and the system should be initialized. The NFSR bits are loaded with 128-bit key. The first 96 bits of the LFSR are loaded with initialization vector (IV) bits. The last bit of the LFSR is loaded with 0 and the remaining bits are loaded with 1.

$$b_i = k_i \qquad \text{for } 0 \le i \le 127 \qquad \text{NFSR Load} \tag{8.8}$$

$$\begin{aligned} s_i &= IV_i && \text{for } 0 \le i \le 95 \\ s_i &= 1 && \text{for } 96 \le i \le 126 && \text{LFSR Load} \\ s_i &= 0 && \text{for } i = 127 \end{aligned} \tag{8.9}$$

After the content of the LFSR and NFSR are loaded, the system is clocked 256 times. This phase is called **initialization** and during this phase the content of y is also XORed with the f and g functions.

Grain-128a algorithm provides two different modes of operation: with and without authentication. This is determined depending on the first value of the IV; if it is 1, authentication is performed; if not, it is forbidden.

8.2.1 Authentication

In the authentication case, two 32 bit registers are used. These are called **accumulator** and **shift register**. Let's denote the content of accumulator by $a_i^0, a_i^1, \ldots, a_i^{31}$ and shift register by $r_i, r_{i+1}, \ldots, r_{i+31}$ at time i. These registers are loaded during the start of the operation, just after initialization. For the first 32 clock cycle, accumulator is loaded by the content of y and for the next 32 clock cycle, shift register is loaded by the content of y according to

$$a_0^j = y_j \qquad \text{for } 0 \le j \le 31 \qquad \text{Accumulator Load} \tag{8.10}$$

$$r_i = y_{32+i} \qquad \text{for } 0 \le i \le 31 \qquad \text{Shift Register Load} \tag{8.11}$$

In case of authentication, the accumulator is updated as

$$a_{i+1}^j = a_i^j + m_i r_{i+j} \tag{8.12}$$

The operation ends with the end of the input message (plain text). The final content of the accumulator is called **tag** and it is used for authentication.

8.2.2 *Key Stream Generation*

For both of the cases (with and without authentication) the key stream is generated but with different ways. The cases are shown below:

$$k_i = \begin{cases} y_{64+2i} & \text{with authentication} \\ y_i & \text{without authentication} \end{cases} \tag{8.13}$$

8.2.3 *Output Rate*

It can be seen from the feedback and output functions that the last 32 bits of LFSR and NFSR are never used. This is intentionally done to provide more options for higher data rates (×2, ×4, ×8, ×16 and ×32). It is possible to increase the number of keystream bits per one cycle by only adding additional feedback functions and using the same shift registers.

8.3 MCML Cell Library

The ×2 option of Grain-128a block was implemented with the semi-custom design approach. The implementation is made with 0.18 μm bulk CMOS technology. The desired clock frequency is set to 780 kHz. For having the optimum design at the given clock frequency, an MCML standard cell library is designed.

8.3.1 *Library Parameters*

In the beginning of the standard cell design, the tail current for the weakest gate (with ×1 drive strength) was calculated to be at least 6.25 nA. After the post layout simulation, it was observed that the minimum current needed for 780 kHz operation is needed to be 6.8 nA with 9% error. Therefore, the tail current for the weakest gate (×1) is set to 6.8 nA, and the swing voltage is set to 250 mV. Finally, a minimum supply voltage of 0.7 V is obtained.

8.3.2 *Cell Selection*

For decreasing the logic depth and number of the gates, some complex logic functions are needed in a single gate. For that, a special standard-cell library is designed which contains different gates with various functions, D-flip flops, and latches. Each gate has six different drive strength options ranging from $\times 1$ to $\times 32$ with the multiples of two. The library contains the following logic functions:

- Inverter/Buffer
- AND2
- XOR2
- 2 to 1 Multiplexer
- AND3
- XOR3
- F9 (XOR2 and AND2 together)
- Latch with Reset
- D Flip Flop with Reset

For having six different drive strengths with a unique bias voltage for the current sources and the loads, it is needed to have six different loads and current sources. The most reliable way to implement the current sources and the loads for the gates with strong drive strengths is to provide multiple finger devices. For example, if the current source in the replica bias circuit has a single device, then the gate with $\times 32$ drive strength option has to have 32 fingers for providing the desired speed. However, this solution increases the size of the strong gates significantly. Therefore, it is decided to have two different replica bias circuits where the first one is uniquely used for the gates with $\times 1$, $\times 2$, $\times 4$ drive strengths and the second one is used for the $\times 8$, $\times 16$, $\times 32$ options. That way, the size of the three strongest gates ($\times 8$, $\times 16$ and $\times 32$) is decreased approximately by 8. Moreover, for the implementation of these gates, the $\times 1$, $\times 2$ and $\times 4$ drive strength gates are used only by changing their current source and load bias voltage inputs.

8.3.3 Cell Characteristics

The noise margin histograms, NMH-NML scatter plot, their standard deviation and mean values are shown for a $\times 4$ buffer/inverter gate in Fig. 8.3a–d. The plotted values are obtained by performing a Monte Carlo simulation of 1000 runs. The swing voltage is set to 250 mV and the tail current is equal to 25 nA. It can be seen from the scatter plot that the worst case noise margin is more than 35 mV. The noise margin performance can also be examined by plotting the butterfly curve of the logic gate and measuring the dimension of the maximum size square that can be placed inside the opening. The butterfly curve for the same logic gate can be seen in Fig. 8.4 where the family of transfer curves are plotted by performing a Monte Carlo simulation of 1000 runs. It can be seen that a square with a side length of 6 mV can be placed inside the openings.

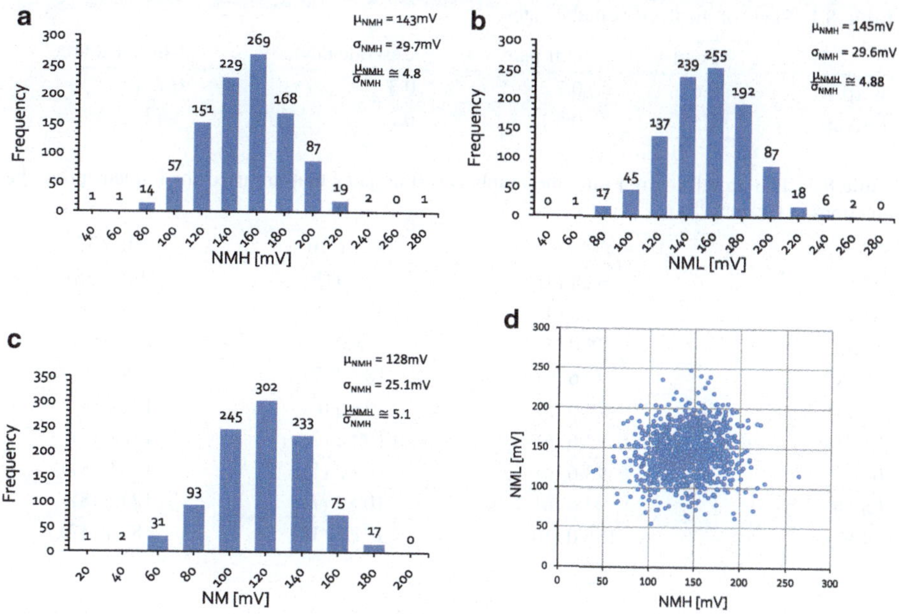

Fig. 8.3 (**a**) NMH histogram, (**b**) NML histogram, (**c**) NM histogram, (**d**) NMH-NML scatter plot

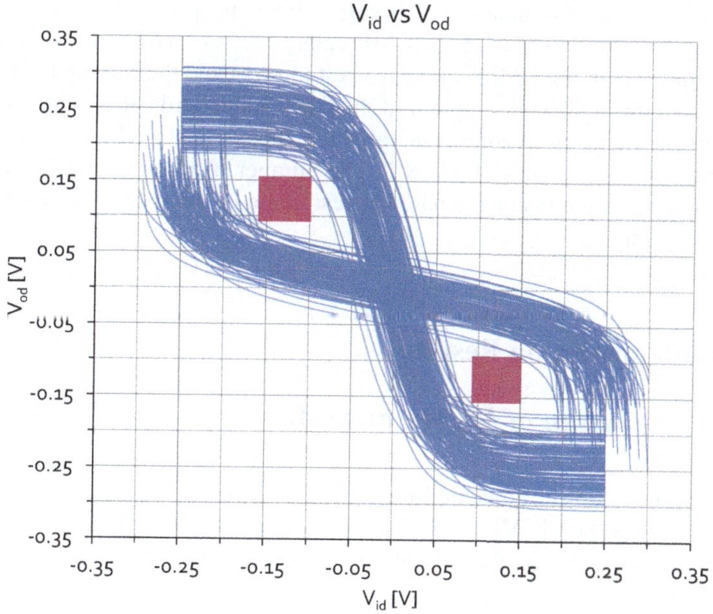

Fig. 8.4 Transfer curves of an inverter to measure the noise margin under variations; 0.18 μm technology, $V_{SW} = 250$ mV, $I_{SS} = 25$ nA

Table 8.1 Sizes of the devices in the gates

	Load	Diff pair	Cascode device	Current source
W (μm)	0.6	0.7	0.3	0.7
L (μm)	1.55	0.3	0.2	1.87

Table 8.2 The area [in μm^2] (and the number) of the gates that are used for implementing the functions in Grain-128a

	×1, ×8	×2, ×16	×4, ×32
BUF	62.4 (4)	78.0 (5)	93.6 (6)
AND2	62.4 (4)	78.0 (5)	109.2 (7)
XOR2	78.0 (5)	78.0 (5)	109.2 (7)
MUX2	93.6 (6)	109.2 (7)	124.8 (8)
AND3	78.0 (5)	93.6 (6)	124.8 (8)
XOR3	93.6 (6)	109.2 (7)	140.4 (9)
F9	93.6 (6)	109.2 (7)	124.8 (8)
Latch	93.6 (6)	109.2 (7)	124.8 (8)
DFF	156.0 (10)	171.5 (11)	187.1 (12)

8.3.4 *Cell Layout*

The layouts of the MCML standard cells are designed compatible to an already existing static CMOS standard-cell library where the pitch of the static CMOS library is used. The height of the MCML cells is set to 13.68 μm (three times the height of the cells in the CMOS standard-cell library) and the routing pitch is set to 1.14 μm (two times the routing pitch of the CMOS standard cells). The sizes of the devices inside the gates are shown in Table 8.1. The area of each cell with different drive options can be seen in Table 8.2. The number of the horizontal pitches that are occupied is written inside the parentheses. Some layout examples can be seen in Figs. 8.5 and 8.6.

8.4 Implementation Results

8.4.1 *Comparison of MCML and CMOS*

After the optimizations of the combinational logic, the logic depth of Grain-128a is decreased to 5 for the MCML implementation. The number of the gates used for the implementations of the feedback functions f, g, the output function h and y can be observed in Table 8.3.

Table 8.4 shows the number of the gates used to implement the core of the Grain-128a which includes the blocks shown in Fig. 8.2 and excludes the authentication part. The gate count summary for the authentication block can be seen in Table 8.5.

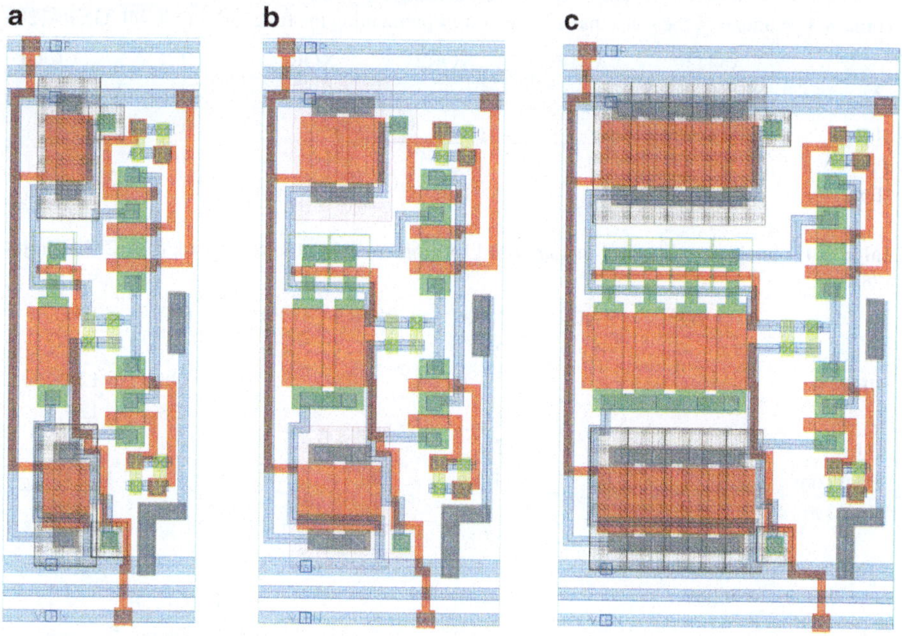

Fig. 8.5 Layouts of AND2 MCML gates for all different drive strengths; 0.18 μm technology. (**a**) ×1 and ×8, (**b**) ×2 and ×16, (**c**) ×4 and ×32

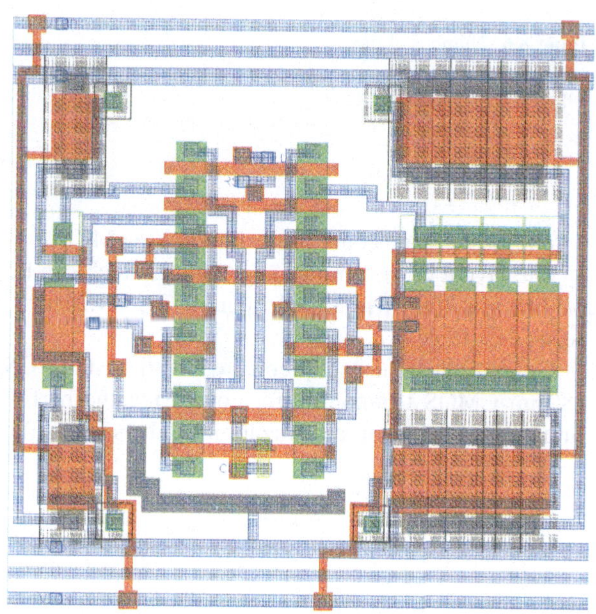

Fig. 8.6 Layout of DFF for ×4 or ×32 drive strengths

Table 8.3 Number of the gates that are used for implementing the functions in MCML Grain-128a

	AND2	AND3	XOR2	XOR3	F9	Total
f		2		1		3
g	4	2	1	4	6	17
h and *y*		1		4	4	9
Total	4	5	1	9	10	**29**

Table 8.4 Number of the gates that are used for implementing the core block of MCML Grain-128a

	AND2	AND3	XOR2	XOR3	F9	DFF	Total
Functions	8	10	2	18	20		58
NFSR						128	128
LFSR						128	128
Other			2		4	4	4
Total comb.	8	10	4	18	24		**64**
Total seq.						260	**260**

Table 8.5 Number of the gates that are used for implementing the authentication block

	F9	DFF
Shift register		32
Accumulator		32
Logic	32	
Total	**32**	**64**

Table 8.6 Number of the gates that are used for implementing Grain-128a

	Logic	DFF
Core	64	260
Authentication	32	64
Total	**96**	**324**

The overall gate count of the Grain-128a MCML block can be seen in Table 8.6.

The layout of the Grain-128a MCML block can be seen in Fig. 8.7. The area of the block is $300.3\,\mu\text{m} \times 264.9\,\mu\text{m} = 79{,}544.2\,\mu\text{m}^2$. The number of the gates can be seen in Table 8.7.

Grain-128a is also implemented with a static CMOS standard-cell library in 180 nm CMOS technology for comparing the performances of two designs. The summary of the resource utilization can be seen in Table 8.7.

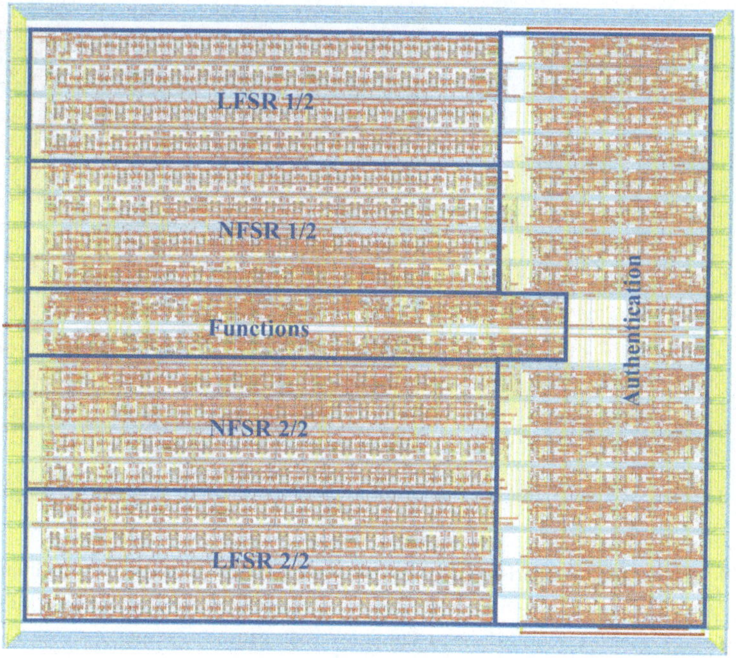

Fig. 8.7 Layout of Grain-128a MCML block with the indicated sub-blocks

Table 8.7 Number of the gates that are used in the CMOS implementation of Grain-128a

	Core only	The rest	Total
BUF		17	17
INV	9	68	77
AND2	2		2
OR2		1	1
NAND2	26	32	58
NAND3	6		6
NAND4	2		2
NOR2		1	1
XNOR2	68	2	70
AND2-OR2		32	32
OR2-NAND2	4		4
2OR2-NAND2		31	31
3OR2-NAND3		32	32
DFF	256	68	324
Comb	**117**	**216**	**333**
Seq	**256**	**68**	**324**

The area of the CMOS implementation is $194.5\,\mu\text{m} \times 194.6\,\mu\text{m} = 37{,}849.7\,\mu\text{m}^2$ and the core density is 81.5%. Figure 8.8 shows both MCML and CMOS implementation layouts together for comparison and Table 8.8 gives a comparison

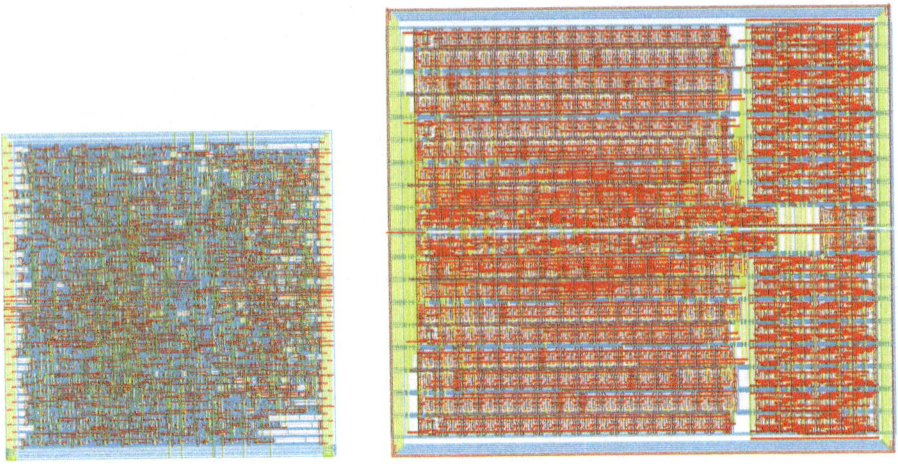

Fig. 8.8 Layouts of CMOS (on the right) and MCML (on the left) implementations of Grain-128a×2

Table 8.8 Performance comparison of the MCML and CMOS implementations of Grain-128a×2

	MCML	CMOS	MCML vs CMOS (%)
Area (μm^2)	79,544	37,849	48
Core density	92	81.5	113
Minimum supply voltage (V)	0.7	0.9	129
Current consumption (μA)	28.7	7.73	27
Power consumption (μW)	20.1	6.96	35
Critical path delay (ns)	1250	2.4	0.2
Frequency (kHz)	800	800	100
Power delay product (pJ)	25.13	8.7	35
Area delay product (mm^2 ns)	99.43	47.31	48
Number of registers	324	324	100
Number of gates	96	333	347
Glitch (Δ current) (μA)	7.5	18,000	240,000

of the MCML and the CMOS implementation of Grain-128a×2. The rightmost column shows how better the MCML implementation is with regard to the CMOS implementation. The area of the MCML implementation is around two times the CMOS implementation. The greater area in MCML implementation is an expected result since the MCML gates occupy four to six times greater area than the CMOS gates. However, the area ratio of the overall implementations is clearly smaller than the area ratios of the MCML and CMOS gates. This is mainly due to the fact that complex MCML gates are capable of decreasing the logic depth and the total number of the gates in an implementation which is a clear result and can also be seen in Table 8.8.

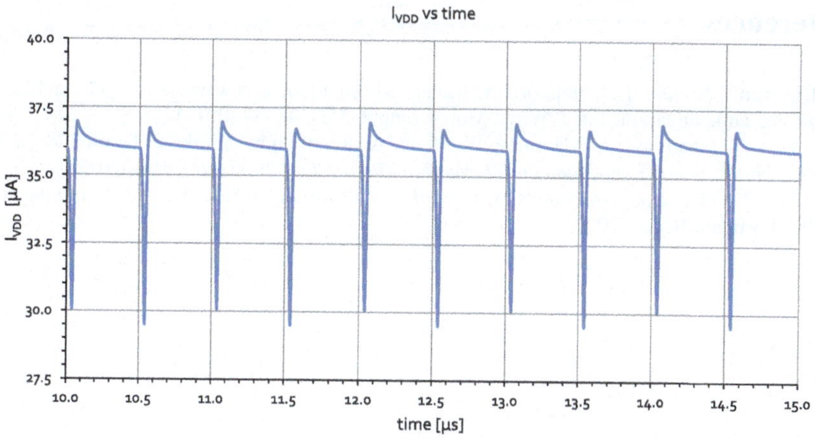

Fig. 8.9 The power supply current consumption waveform of the MCML implementation of Grain-128a×2

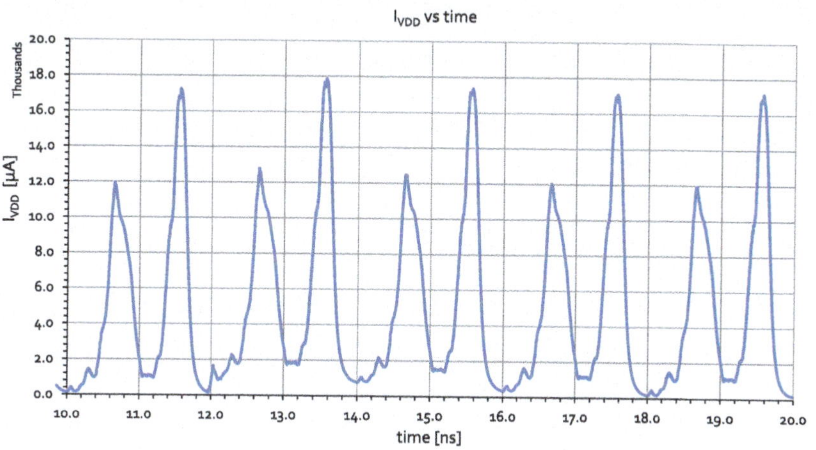

Fig. 8.10 The power supply current consumption waveform of the CMOS implementation of Grain-128a×2

Finally, the power supply current consumption of the MCML and the CMOS implementations of Grain-128a×2 can be seen in Figs. 8.9 and 8.10. It can be seen that the difference between the maximum and the minimum points on the current consumption waveform is around 7.5 µA for the MCML implementation. Whereas in the CMOS implementation, the difference of the peaks is 24 mA. It can seen that the current variations between MCML and CMOS cases are quite different which show the superiority of MCML against CMOS in terms of hiding the signature and the activity of cryptographic circuits.

References

1. M. Ågren, M. Hell, T. Johansson, W. Meier, Grain-128a: a new version of Grain-128 with optional authentication. Int. J. Wirel. Mob. Comput. **5**(1), 48–59 (2011)
2. M. Hell, T. Johansson, A. Maximov, W. Meier, A stream cipher proposal: Grain-128, in *2006 IEEE International Symposium on Information Theory*, July 2006, pp. 1614–1618
3. C. Paar, J. Pelzl, *Understanding Cryptography: A Textbook for Students and Practitioners*, 1st edn. (Springer, Berlin, 2009)

Chapter 9
Design Example IV: Advanced Encryption Standard (AES)

9.1 Circuit Description

MOS Current Mode Logic (MCML) is one of the most promising logic styles to counteract power analysis attacks as it has been already discussed in the previous chapter. Unfortunately, the static power consumption of MCML standard cells is significantly higher compared to equivalent functions implemented using static CMOS logic. As a result, the use of such a logic style is very limited in portable devices. Paradoxically, these devices are the most sensitive to physical attacks, thus the ones which would benefit more from the adoption of MCML. To overcome this limitation, the static power consumption of MCML-based cryptographic circuits must be drastically reduced in these devices. Interestingly, very often, cryptographic functions included in embedded devices are inactive over a long period of time, hence they can be switched off when there is no activity. Power Gated MCML (PG-MCML), a technique for implementing MCML standard cells which contain a sleep transistor in every cell, is an appropriate candidate for implementing low power consumption cryptographic blocks in MCML. Therefore a cryptographic block implementation by using a PG-MCML library might provide similar power consumption performance as CMOS while having a more robust performance in terms of security.

Some examples of PG-MCML can be found in the literature. An appealing target for PG-MCML is represented by the DPA-resistant instruction set extension (ISE) proposed in [6, 7]. In these works, the authors proposed to augment a processor, realized with conventional standard cell libraries, with additional functional units (in the form of custom instructions) implemented in a logic style robust against power analysis attacks. Considering the relevance of such an example, the same approach is used to evaluate the power consumption of PG-MCML.

During this implementation, a software implementation of the AES algorithm [5] is used. The OpenRISC 1000 [4] 32-bit embedded processor with a custom functional unit is designed, sitting in the processor's pipeline, consisting of four

S. Badel et al., *Design Automation for Differential MOS Current-Mode Logic Circuits*, https://doi.org/10.1007/978-3-319-91307-0_9

identical S-boxes (each S-box is implemented in the form of 8 × 8 look-up-table) to match the processor's word size. The new custom instruction is labeled S-box ISE. Both the processor and the custom instruction are available as RTL code. The RTL code is synthesized, placed, and routed three different versions of the considered core. In all cases, the processor was realized using the reference static CMOS technology, a 90 nm commercial standard cell library, while the protected instruction was implemented in three different cases which are (1) using the same conventional CMOS technology, (2) the conventional MCML, and (3) the PG-MCML, respectively.

The custom functional unit implemented with differential cells was connected to the processor by means of voltage level converters and it appears in the processor's layout as a macro block. The full design was synthesized, placed, and routed setting 400 MHz as operating frequency (to meet the speed requirement of modern embedded systems).

9.2 MCML Cell Library

The MCML cell library implemented for this design and described in this section targets a 90 nm CMOS process, with 1.0 V supply voltage. The library is designed by including the power gating property to the MCML cells.

9.2.1 Standard Cell Design with Power Gating

Power gating [3] is a technique which reduces the static power consumption of a digital circuit by inserting power switches (sometimes referred to as sleep transistors) in the supply path. To implement this technique, two solutions have been proposed in the past: *coarse-grain power gating*, in which complete blocks are disconnected from the power supply and the ground through a common power switch, and *fine-grain power gating*, in which every standard cell contains a sleep transistor internally.

In conventional static CMOS circuits, the use of fine-grain power gating causes a significant area overhead and negatively affects performance. For this reason, coarse-grain power gating is the preferred approach for static CMOS. On the contrary, the insertion of a sleep transistor in each MCML cell introduces negligible power overhead. Also, the switching speed is not directly affected by the sleep transistor since it is located outside the signal path. Therefore, in the designed library, fine-grain power gating is implemented, which suits better the needs of MCML cells. Moreover, a fine-grain power gating allows to selectively switch off each standard cells depending on the circuit topology: this step can be easily

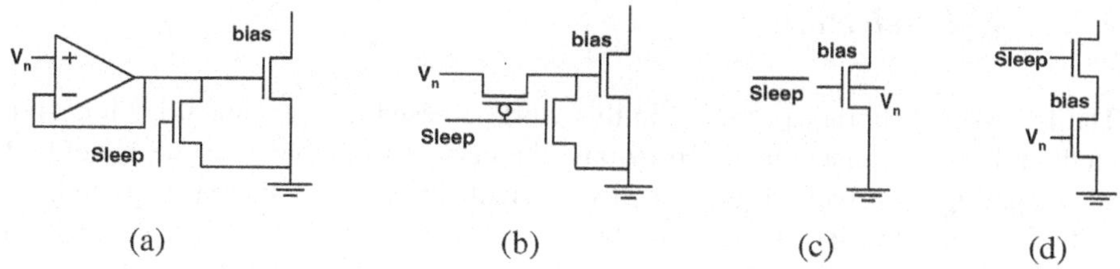

Fig. 9.1 Different power gating techniques for MCML circuits

Table 9.1 Area comparison between conventional MCML and PG-MCML standard cells in 90 nm CMOS technology

Cell	MCML (μm^2)	PG-MCML (μm^2)
BUFX1	7.056	7.448
MUX4X1	19.7568	20.8544
AND4X1	16.9344	17.8752
DLX1	8.4672	8.9376

automated during the synthesis process, using an approach similar to automatic clock gating. However, it should be taken into account that many of the existing synthesis tools currently do not offer the capabilities of fine-grain power gating.

Different power gating topologies for MCML standard cells are depicted in Fig. 9.1. The solutions (a) and (b) use a transistor to pull down the bias voltage V_n to ground during the sleep mode. Solution (c) applies just an ON signal to the gate of the current source and connects the bulk voltage to the bias voltage V_n. Option (d) consists of an additional sleep transistor in series with the current source.

Indeed, solution (a) was discarded since it requires the use of a large bandwidth source follower amplifier to settle the output voltage to V_n within a single clock cycle. Option (b) slightly improves option (a). However, this solution was discarded too, since it requires the insertion of two transistors per cell. Solution (c) relies on the body-biasing principle and modulates the bias current adjusting the threshold voltage. However, to ensure a correct functionality in all the process corners, the voltage V_n needs to range from −500 mV to 1 V. Such voltage is difficult to obtain in practice. In addition, the current source is sitting in a separate well. This solution leads to a significant area overhead. For all the above reasons, solution (d) is selected. It can be seen from Fig. 9.1 that the sleep transistor is located on top of the current source. Thanks to this choice, the sleep transistor has a negative V_{GS} voltage during power down, decreasing the leakage current. Due to its superiority against the other solutions, solution (d) is used for implementing the PG-MCML standard cell library. Table 9.1 shows the silicon area of MCML gates with and without sleep transistor. On average, the cells with sleep transistor are approximately 6% larger than conventional MCML gates.

9.2.2 Cell Selection

The PG-MCML library proposed in this work is based on the standard cell design methodology explained in the previous chapters. To demonstrate the benefit of power gating on a real circuit a relatively small library is designed, including 16 cells including the latches and the buffers. Nevertheless, it is worth to mention that an increased number of cells would positively affect the results because of the higher flexibility offered during synthesis, placement, and routing. The list of the implemented logic functions is listed below:

- Inverter/Buffer
- Differential to Single Ended Converter
- 2-Input AND
- 3-Input AND
- 4-Input AND
- 2 to 1 Multiplexer
- 4 to 1 Multiplexer
- 3-Input Majority
- 2-Input XOR
- 3-Input XOR
- 4-Input XOR
- D-Latch
- D Flip Flop
- D Flip Flop with Reset Input
- D Flip Flop with Enable Input
- Full Adder

The internally developed PG-MCML library is specifically optimized for area and power, and the switching speed of the PG-MCML cells is similar to the one of their CMOS counterparts implemented on the same technology. As previously mentioned, the main target application is security, with particular emphasis on battery operated embedded systems which need to be robust against power analysis attacks.

The main difference between an MCML standard cell which includes a sleep transistor and its conventional counterpart is that, in the former, the minimal supply voltage and the current source are slightly increased. Finally, to minimize the layout area, the sleep transistor and the current source are sized with the same channel width to share the same diffusion region.

9.2.3 Cell Characteristics

The PG-MCML library is designed using a 90 nm CMOS process. To improve timing, area and power each cell is implemented using a combination of low-V_t and high-V_t transistors. Indeed, high-V_t devices can reduce the leakage current during

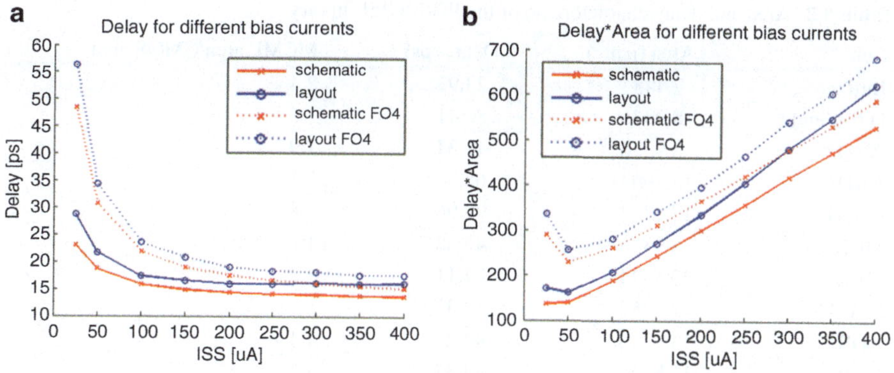

Fig. 9.2 Delay and area delay trade-off for an MCML inverter driving FO1 and FO4 loads

sleep mode without affecting the cell delay, thus high-V_t devices are selected for the NMOS Boolean network, the current source and the sleep transistor. Low-V_t devices are used for the PMOS load, since this approach leads to the smallest silicon area. Furthermore, smaller active loads have less parasitic and thus they lead to higher speed. To determine the optimal bias current I_{SS}, it is explored how the cell delay and the power consumption vary in function of the tail current. Figure 9.2a depicts the delay for an MCML buffer/inverter driving FO1 and FO4 loads. Interestingly, increasing the bias current above 250 μA provides a limited speed improvement with a large penalty in cell area. Figure 9.2b depicts the power delay product under different bias conditions. The simulation based evaluation revealed a minimum area delay product at 50 μA.

Table 9.2 shows area and delay for the cells belonging to the PG-MCML library. An area comparison between equivalent cells in a commercial 90 nm standard cell library is also provided. PG-MCML cells are 1.6 times larger in average. To the best of our knowledge, this is the smallest area overhead measured for a power analysis resistant library so far. Figure 9.3 depicts the schematic and the layout view of a buffer cell belonging to the PG-MCML library. Both driving strengths one and four are shown.

The cells are designed to support fine-grain power gating. Theoretically, in a design, there are several power gating opportunities which could be detected automatically and exploited during synthesis. However, since modern synthesis tools are specifically designed for conventional CMOS libraries, they do not support fine-grain power gating. Due to this limitation, the sleep transistor is manually connected. For fine-grained power gating applications, each individual sleep input of all cells in a cluster can be driven collectively, provided that the signal is sufficiently buffered.

The design flow used for PG-MCML is based on commodity EDA tools for synthesis, placement, and routing. *Synopsys Design Compiler* is used for synthesis and *Cadence Encounter Digital Implementation* for placement and routing. The last step exploits the fat-wire approach [1], which ensures that both wires of a

Table 9.2 Area and delay characteristic of the PG-MCML library

Cell	Area (μm^2)	Delay (ps)	MCML area/CMOS area
Buffer	7.448	23.97	2.4
Diff2Single	8.9376	80.41	
AND2	8.9376	41.34	1.9
AND3	13.40641	68.74	2.1
AND4	17.8752	99.96	2.8
MUX2	8.9376	43.58	1.2
MUX4	20.8544	87.11	1.2
MAJ32	17.8752	82.32	
XOR2	8.9376	44.26	1.1
XOR3	17.8752	84.37	1.1
XOR4	20.8544	109.68	1.1
D-Latch	8.9376	36.32	1.3
DFF	17.8752	53.4	1.3
DFFR	26.8128	69.33	1.8
EDFF	23.8336	63.53	
FA	35.7504	84.49	1.4

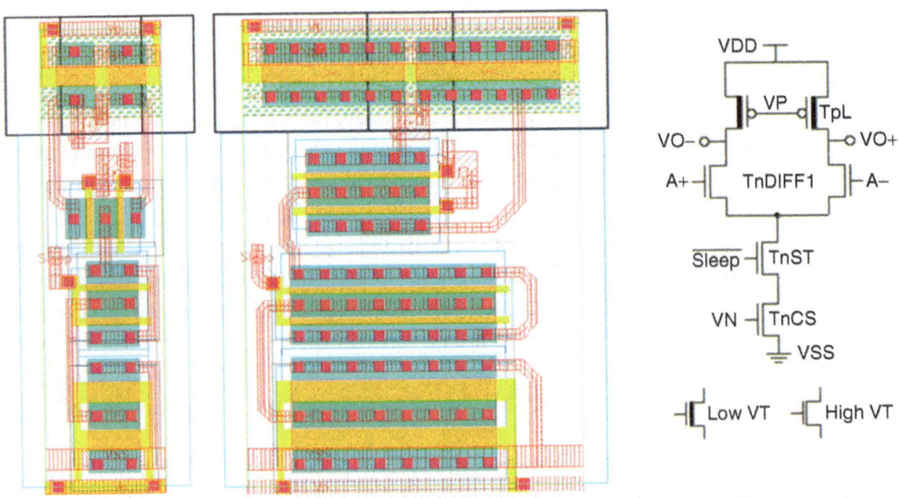

Fig. 9.3 Schematic and layout views of a PG-MCML buffer with drive strength X1 and X4. The bias transistor is laid out close to the ground rail while PMOS load transistors are placed below the power rail. The sleep transistor is located next to the bias transistor for minimal silicon area. Intra-cell routing is limited to Metal-1 to facilitate inter cell routing using upper metal layers

differential signal are routed side by side (to have the same delay and load). In order to switch off and on the controlled logic in a fraction of the clock cycle (in the order of few ns), the sleep signal, managing the power gating, should be buffered.

In practice, to control skew and propagation delay from the root to the individual sleep input of all cells in a cluster, the sleep signal is routed and buffered as a balanced tree. Also, single ended clock buffers are used to route it with a controllable insertion delay. To integrate such cells with the PG-MCML library, static CMOS buffer/inverters with the same height as the PG-MCML cells are designed and characterized. The *clock tree synthesis (CTS)* engine available in the place and route tool is used to synthesize the buffer tree and route the sleep signal. In this way, it would be possible to exploit capabilities already existing in digital design tools.

9.3 Implementation Results

Detailed simulations were performed for observing the ability of the implementations to hide the signature during cryptographic operation. For that, a software implementing the AES cipher was repeatedly executed 5000 times using a random plain-text. The full AES algorithm was simulated with a logic simulator (*Mentor Graphics Modelsim*) using the post place and route netlist and the delay back annotation (in SDF format) as input.

For such a benchmark, the S-box ISE under evaluation was active 0.01% of the whole execution time. The signal triggering the custom instruction's execution controls also the sleep signal, so that the protected logic is turned on only during the custom instruction execution. The sleep signal is shared among all the cells that compose the custom instruction and its insertion delay is approximately 1 ns. This allows to turn on the custom instruction in a small fraction of the clock period and to process the data within the same cycle. The circuit's functionality has been verified in simulation.

The custom instruction's inputs, stored in VCD format, are then used to run transistor level simulations (using *Synopsys Nanosim* as fast SPICE simulator) to monitor the custom instruction's current consumption. Figure 9.4 depicts the current waveform of the S-box ISE realized with conventional MCML (dashed line) and the one realized with the designed PG-MCML library (solid line). The clock and sleep signals are also depicted. It can be seen how the power gating allows to significantly decrease the power consumption: the current drawn by the conventional MCML circuit is always flat (around 30 mA), while the current absorption of the power gated S-box ISE is almost negligible when encryption is not performed (sleep signal low).

The overall results are summarized in Table 9.3, which reports the area, the gate occupation, the delay, and the power consumption for the considered S-box instruction set extension implemented in three different logic styles. The average power consumption of PG-MCML is significantly lower compared to the one of conventional MCML (reduced by a factor or 10^4). Also, it can be noticed that PG-MCML consumes four times less power than CMOS. This data should not suggest that PG-MCML is less power hungry than static CMOS logic, since when power gating techniques are applied to CMOS, the power consumption of this

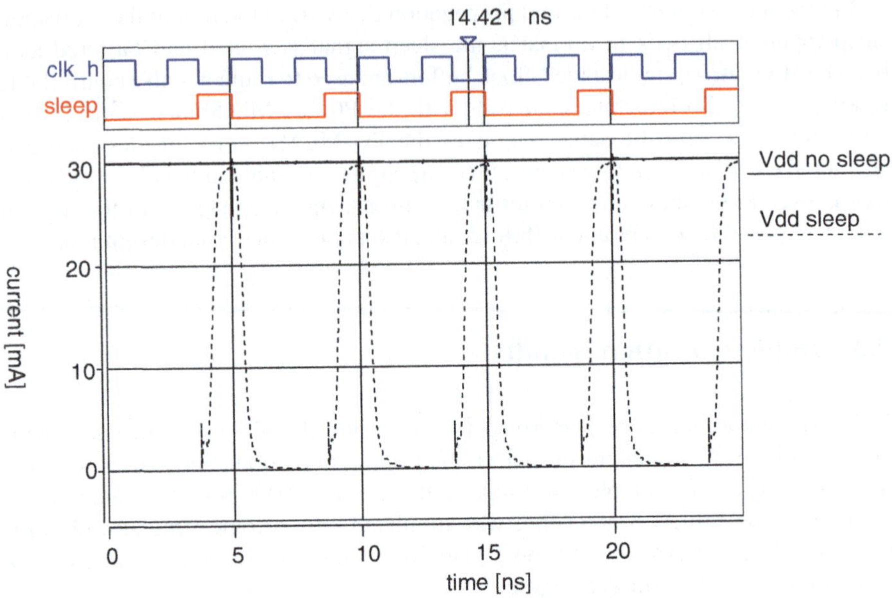

Fig. 9.4 Current waveform for S-box ISE with and without power gating implemented. The sleep signal waveform for the power gated implementation is also plotted

Table 9.3 Area, delay, and power consumption on the S-box ISE implemented in different logic styles

	CMOS	MCML	PG-MCML
Cells	3865	2911	3076
Area (μm^2)	30,547.52	77,378.97	78,355.21
Delay (ns)	0.630	0.698	0.717
Avg power (W)	207.72×10^{-6}	490.56×10^{-3}	47.77×10^{-6}

technology is significantly reduced. The fact that the power consumed by PG-MCML is comparable to the one of static CMOS without power gating is still important, since it proves that when power gating is used, MCML represents an appealing option. Finally, it can be seen that the area overhead necessary to support the sleep signal is negligible (the PG-MCML is roughly 1000 μm^2 larger compared to conventional MCML).

Considering the specific application field, the robustness against power analysis attacks of the PG-MCML library is also carefully evaluated. To evaluate the security, the commonly accepted reduced version of the AES algorithm composed by a key addition and an S-box look-up-table was synthesized, placed, and routed. Each implementation was realized using three different technologies: the reference static CMOS, the conventional MCML, and the PG-MCML. For all of them, SPICE simulations are performed to extract the instantaneous current of all possible plain-text secret key pairs, using very high resolution both for current (1 μA) and time

(1 ps). Finally, all the implementations were repeatedly attacked using as power model the Hamming weight of the S-box output [2]. More in detail, a typical power analysis attack [2] consists of selecting as target an intermediate key dependent result, encrypting a certain number of known plain-texts and measuring the corresponding power consumption, calculating hypothetical intermediate values based on a key guess and evaluating the guess over the power consumption. The correlation between the collected power traces and the various key hypothesis will show a peak in correspondence to the correct key, while for all the other key guesses it is close to zero.

As expected, all the attacks on the CMOS implementations were successful, while none of the ones performed on conventional MCML as well as on PG-MCML were able to reveal the secret key. In fact, as it can be seen from Fig. 9.5, which reports an example of correlation power attack (CPA) on PG-MCML, the secret key, plotted in black, is not distinguishable from all the other key guesses plotted in light gray. The experiments showed that the security level achievable using the proposed PG-MCML is comparable to the conventional MCML, thus the insertion of the sleep signal does not introduce a negative effect on robustness against power analysis attacks.

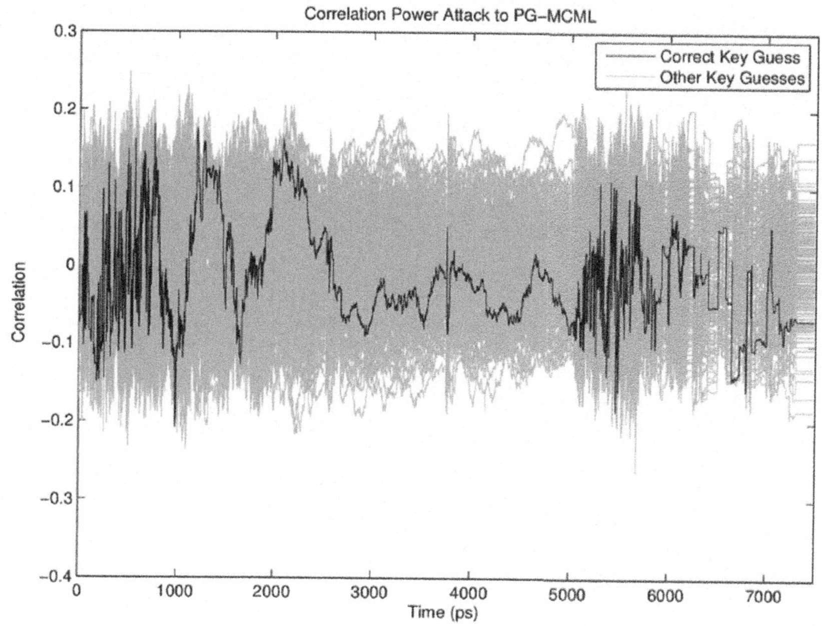

Fig. 9.5 Correlation power attacks to PG-MCML: the black line, corresponding to the secret key, is not distinguishable

References

1. S. Badel, E. Güleyüpoğlu, Ö. İnać, A. Peña Martinez, P. Vietti, F.K. Gürkaynak, Y. Leblebici, A generic standard cell design methodology for differential circuit styles, in *Design Automation and Test in Europe (DATE)*, 2008
2. E. Brier, C. Clavier, F. Olivier, Correlation power analysis with a leakage model, in *Cryptographic Hardware and Embedded Systems—CHES 2004*, ed. by M. Joye, J.-J. Quisquater. Lecture Notes in Computer Science, vol. 3156 (Springer, Berlin, 2004), pp. 16–29
3. S. Henzler, *Power Management of Digital Circuits in Deep Sub-micron CMOS Technologies* (Springer, Dordrecht, 2006)
4. D. Lampret, OpenRISC 1000 Architecture Manual, April 2006
5. NIST, Announcing the Advanced Encryption Standard (AES). Federal Information Processing Standards Publication 197, November 2001
6. F. Regazzoni, A. Cevrero, F.-X. Standaert, S. Badel, T. Kluter, P. Brisk, Y. Leblebici, P. Ienne, A design flow and evaluation framework for DPA-resistant instruction set extensions. In: CHES '09 Proceedings of the 11th International Workshop on Cryptographic Hardware and Embedded Systems, Lausanne, September 2009
7. S. Tillich, J. Großschädl, Power analysis resistant AES implementation with instruction set extensions. In: Proceedings of the 9th International Workshop on Cryptographic Hardware and Embedded Systems, vol. 4727 (Springer, Berlin, 2007), pp. 303–319

Chapter 10
Conclusions

The main goal of this work has been to investigate the implementation of complete digital blocks in MOS Current-Mode Logic (MCML) with a standard-cell methodology, using existing frameworks as much as possible.

Though MCML had been studied previously, one contribution of this work is a new, extended, analytical model for MCML logic gates that takes into account the weak-inversion behavior of MOS transistors. This model is shown to be more accurate than existing models, and fundamental limits of the voltage swing due to weak-inversion behavior are derived.

In the aim of developing a standard-cell library in MCML, a systematic analysis of MCML logic gates based on a physical analogy with binary decision diagrams has been presented, and used to derive an exhaustive list of possible MCML gate implementation up to three levels of stacked logic stages.

From the library to the implementation, several issues are addressed to adapt the existing automation tools and design flow to suit the specificities of MCML circuits. Specifically, the differential nature of the signals is hardly supported by any existing tools for digital IC implementation, and part of this work is devoted to bringing solutions to enable to take full advantage of differential signaling in an automated way. This includes the creation of a number of custom automation tools to fill the gaps during the library characterization, logic synthesis, and placement and routing steps. These tools were created with the hope that they could be reused, thus a particular accent has been put on keeping them as generic and easy-to-use as possible.

Finally, an implementation of a library of MCML standard cells in a 0.18 μm CMOS technology is presented. Using the proposed design methodology, this library is used to redesign an existing CMOS digital block devoted to analog-to-digital applications, a mixed-signal environment where the low-noise properties of MCML cells potentially represent an important benefit.

S. Badel et al., *Design Automation for Differential MOS Current-Mode Logic Circuits*, https://doi.org/10.1007/978-3-319-91307-0_10

The resulting circuits are compared to the original implementation, and it is shown that, while the MCML circuit consumes more power, it is able to achieve the target clock frequency of 1GHz, requiring little more silicon area and reducing the generated supply noise by more than one order of magnitude.

10.1 Future Work

The higher power consumption of the MCML logic style certainly represents an obstacle to its use in highly complex logic circuits. The static power consumption in each gate renders MCML circuits much less power-efficient than conventional CMOS circuits especially when the activity factor is low.

Several areas could be investigated in order to reduce the power consumption of MCML circuits. First, the use of lower threshold voltages has been proved to allow an increase of the speed in MCML circuits. The use of lower threshold voltages could also enable lowering of the supply voltage, allowing to substantially decrease the power consumption. One main issue when scaling the supply voltage is the linearity of the pull-up devices, and investigating new types of load devices could be an interesting area of research.

Another approach to reduce the power consumption is dynamic power management. At the circuit level, dynamic voltage and frequency scaling is an efficient technique to reduce the power consumption by reducing the operating speed of parts of a circuit when full-speed operation is not necessary. When the operating speed is reduced, the power supply voltage can also be reduced, resulting in substantial power savings. However, in classical circuits, the generation of several supply voltages necessitates the implementation of costly on-chip voltage regulators. In contrast, the power-speed tradeoff in MCML circuits is inherently and easily controllable through the bias voltages: both speed and power vary linearly with the bias current. With the controllable current consumption, the supply voltage can be kept constant, eliminating the need for voltage regulation. Thus, the integration of dynamic power management units in MCML circuits could be an attractive way to enhance the power efficiency.

Another interesting area of application for MCML is cryptographic circuits. Differential power attack (DPA) is an efficient way to discover secret information in secure circuits, and thanks to their superior noise performance, MCML gates inherently leak few information about the state of the circuit through the power supply. Yet, MCML circuits could certainly be made more secure by a deeper understanding of the noise production mechanisms and appropriate counter-measures.

Appendix A
Large-Signal Transitional Model of the MOS Differential Pair

In most of the cases, the large-signal operation of the differential pair cannot be analyzed accurately by considering a single mode of operation for the transistors—the exception being when it is operated in deep weak-inversion, over a range of currents where the subthreshold exponential $I_D - V_G$ relationship holds with good accuracy. When biased in strong (or moderate) inversion, the strong-inversion model will only remain valid as long as the gate-to-source voltages of both transistors remain large enough compared to V_T, or equivalently as long as the current remains high enough in both branches. This is the reason why a strong-inversion model can correctly predict the behavior of the differential pair in the central region of the transfer characteristic, which is close to linear, but fails to accurately model the regions where the curve is bending.

Modern transistor models provide continuous, transregional equations for the drain currents. In particular, in the EKV MOSFET model, the weak- and strong-inversion regimes are considered separately, and a continuous expression is derived by defining an inversion coefficient, which reflects the operating state of the device, and using a smooth interpolation function of this coefficient which asymptotically tends towards the weak- and strong-inversion expressions, respectively, when $I_D \longrightarrow 0$ and when $I_D \longrightarrow \infty$. We will use the same approach to bridge two expressions in weak- and strong-inversion.

Let us assume that, using a given model for a transistor in strong inversion, the transconductances can be expressed as a function of the drain current as $g_{m(strong)} = g_s(I_D)$. Furthermore, let us assume that, in weak inversion, the transconductance is given by $g_{m(weak)} = g_w(I_D)$. Now let us build a continuous expression for the transconductance which asymptotically tends toward g_w when $I_D \to 0$, and to g_s when $I_D \to \infty$, which is shown by (2.18). This interpolation is valid under the assumptions that $g_w/g_s \to \infty$ when $I_D \to \infty$, and $g_w/g_s \to 0$ when $I_D \to 0$, which hold for MOSFETS. This is graphically represented on Fig. 2.5.

Then, since $g_m = \mathrm{d}I_D/\mathrm{d}V_G$, we can obtain V_G by integrating $1/g_m$ according to (2.19). Now let us calculate the $V_{id} - \Delta I$ relationship for the differential pair using this expression of V_G

S. Badel et al., *Design Automation for Differential MOS Current-Mode Logic Circuits*, https://doi.org/10.1007/978-3-319-91307-0

$$\begin{aligned} V_{id} &= V_G(I_1) - V_G(I_2) \\ &= \left[V_{G(strong)}(I_1) - V_{G(strong)}(I_2)\right] + \left[V_{G(weak)}(I_1) - V_{G(weak)}(I_2)\right] \\ &= V_{id(strong)}(\Delta I) + V_{id(weak)}(\Delta I) \end{aligned} \tag{A.1}$$

where $V_{id(strong)}$ and $V_{id(weak)}$ are the $V_{id} - \Delta I$ transfer curves for the differential pair in strong and weak inversion, respectively. The overall transconductance $g_{md}$is therefore given by

$$\begin{aligned} g_{md} &= \left(\frac{\partial V_{id}}{\partial \Delta I}\right)^{-1} \\ &= \left(\frac{1}{g_{md(strong)}} + \frac{1}{g_{md(weak)}}\right)^{-1} \\ &= \frac{g_{md(weak)} \cdot g_{md(strong)}}{g_{md(weak)} + g_{md(strong)}} \end{aligned} \tag{A.2}$$

At equilibrium,

$$g_{md,0} = \frac{1}{\frac{1}{g_{md,0(weak)}} + \frac{1}{g_{md,0(strong)}}} \tag{A.3}$$

With this definition, we can normalize the relationship

$$V_{id}\frac{g_{md,0}}{I_{SS}} = (1-\delta)\cdot\left[V_{id(strong)}\frac{g_{md,0(strong)}}{I_{SS}}\right] + \delta\cdot\left[V_{id(weak)}\frac{g_{md,0(weak)}}{I_{SS}}\right] \tag{A.4}$$

where

$$\delta = \frac{g_{md,0(strong)}}{g_{md,0(weak)} + g_{md,0(strong)}} = \frac{g_{md,0}}{g_{md,0(weak)}} \tag{A.5}$$

Note

The constant C in (2.19) is chosen for an arbitrary couple of values (V_G, I_D). In particular, if we choose the point where the gate voltages are equal in strong and weak inversion, then $C = -V_T - nV_S$ and

$$\frac{V_G - V_T}{n} - V_S = \left[\frac{V_G - V_T}{n} - V_S\right]_{(strong)} + \left[\frac{V_G - V_T}{n} - V_S\right]_{(weak)} \tag{A.6}$$

Example

Let us build a continuous model using the results from Sects. 2.2.1 and 2.2.2, that is, assuming an ideal strong-inversion behavior. In strong inversion,

$$\left[V_{id}\frac{g_{md,0}}{I_{SS}}\right]_{(strong)} = \sqrt{1+\frac{\Delta I}{I_{SS}}} - \sqrt{1-\frac{\Delta I}{I_{SS}}}$$

$$\left[g_{md,0}\right]_{(strong)} = \sqrt{\frac{\beta I_{SS}}{n}}$$

Similarly, in weak inversion

$$\left[V_{id}\frac{g_{md,0}}{I_{SS}}\right]_{(weak)} = \tanh^{-1}\left(\frac{\Delta I}{I_{SS}}\right)$$

$$\left[g_{md,0}\right]_{(weak)} = I_{SS}/\left(2\cdot nU_T\right)$$

Therefore,

$$g_{md,0} = \frac{I_{SS}}{2nU_T}\frac{1}{1+\sqrt{\frac{I_{SS}/2}{2n\beta U_T^2}}}$$

$$\delta = \frac{1}{1+\sqrt{\frac{I_{SS}/2}{2n\beta U_T^2}}}$$

And the transitional formula is given by

$$V_{id}\frac{g_{md,0}}{I_{SS}} = (1-\delta)\cdot\left(\sqrt{1+\frac{\Delta I}{I_{SS}}} - \sqrt{1-\frac{\Delta I}{I_{SS}}}\right) + \delta\cdot\tanh^{-1}\left(\frac{\Delta I}{I_{SS}}\right)$$

Appendix B
List of MCML Templates up to Three Levels

In the following pages, an exhaustive list of MCML gate topologies with up to three levels of stacked differential pairs is presented. The topologies listed here are sufficient for efficiently implementing a large class of circuits.

This represents a practical reference for designers to select the proper topology in order to implement a given logic function. In this guide, each different physical topology (footprint) is presented on a separate page. The BDD representation of the footprint is displayed, with generic signal names $A_0 - A_N$ assigned to each node. Then, a table lists the templates (implementing an NPN class of logic functions) that can be realized by different mappings of signals onto the differential pair inputs.

We have tried to identify common functions such as AND and XOR, as well as their combinations, in order to provide meaningful names and descriptions for the various templates. Whenever this wasn't possible, an automatic name such as $T4\_11$ has been assigned, where the first number represents the number of inputs to the logic function, and the second number is incremented to produce a unique name.

S. Badel et al., *Design Automation for Differential MOS Current-Mode Logic Circuits*, https://doi.org/10.1007/978-3-319-91307-0

Footprint F1

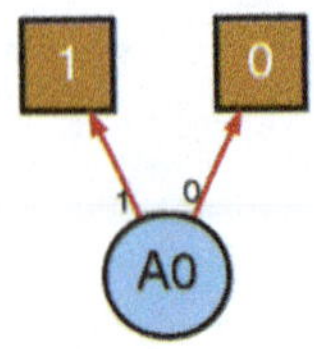

Template BUF

Description: Buffer

Function: $Y = A$

Variants: None

Inputs	Output
$A \Rightarrow A_0$	Y

Footprint F2

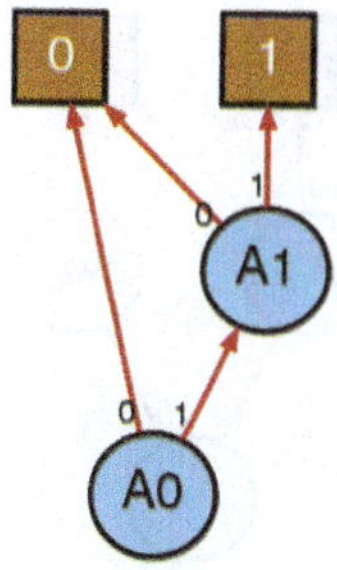

Template AND2

Description: 2-Input AND Gate

Function: $Y = A \cdot B$

Variants: None

Inputs	Output
$A \Rightarrow A_0 \quad B \Rightarrow A_1$	Y

Footprint F3

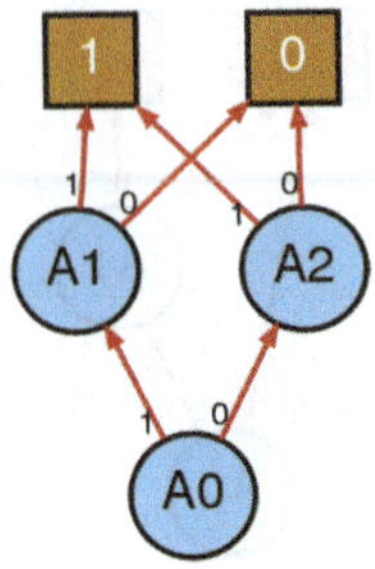

Template XOR2

Description: 2-Input Exclusive-OR Gate

Function: $Y = A \oplus B$

Variants: None

Inputs	Output
$A \Rightarrow A_0$ $B \Rightarrow A_1$ $B \Rightarrow \overline{A_2}$	$\overline{Y}$

Template MUX2V1

Description: 2-to-1 Multiplexer

Function: $Y = S \cdot D_1 + \overline{S} \cdot D_0$

Variants: MUX2V2, MUX2V3

Inputs	Output
$D_0 \Rightarrow A_1$ $D_1 \Rightarrow A_2$ $S \Rightarrow \overline{A_0}$	Y

Footprint F4

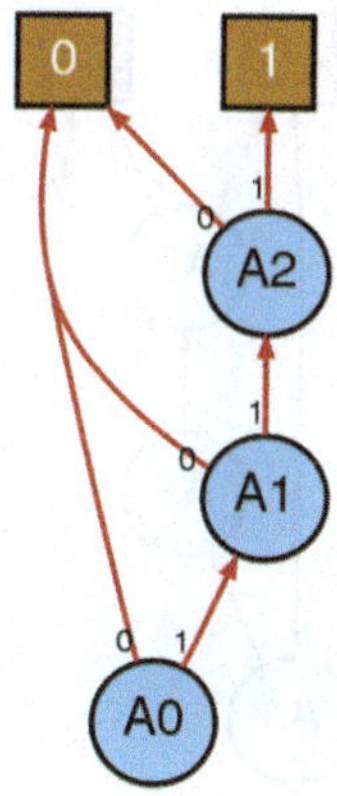

Template AND3

Description: 3-Input AND Gate

Function: $Y = A \cdot B \cdot C$

Variants: None

Inputs	Output
$A \Rightarrow A_0 \quad B \Rightarrow A_1 \quad C \Rightarrow A_2$	Y

Footprint F5

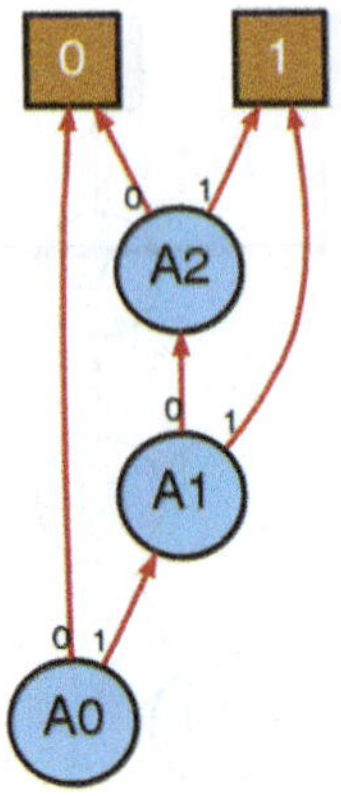

Template AO21V1

Description: 2-Input AND into 2-Input OR Gate

Function: $Y = A \cdot B + C$

Variants: AO21V2, AO21V3

Inputs	Output
$A \Rightarrow \overline{A_1}$ $B \Rightarrow \overline{A_2}$ $C \Rightarrow \overline{A_0}$	$\overline{Y}$

Footprint F6

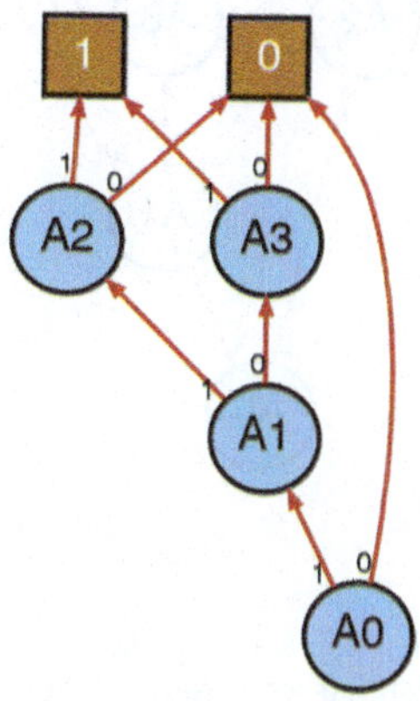

Template XA21V1

Description: 2-Input Exclusive-OR into 2-Input AND Gate

Function: $Y = (A \oplus B) \cdot C$

Variants: XA21V2, XA21V3

Inputs	Output
$A \Rightarrow \overline{A_1}$ $B \Rightarrow A_2$ $B \Rightarrow \overline{A_3}$ $C \Rightarrow A_0$	Y

Template MA21V1

Description: 2-to-1 Multiplexer into 2-Input AND Gate

Function: $Y = A \cdot (S \cdot D_1 + \overline{S} \cdot D_0)$

Variants: MA21V2, MA21V3, MA21V4, MA21V5

Inputs	Output
$A \Rightarrow A_0$ $D_0 \Rightarrow A_2$ $D_1 \Rightarrow A_3$ $S \Rightarrow \overline{A_1}$	Y

Footprint F7

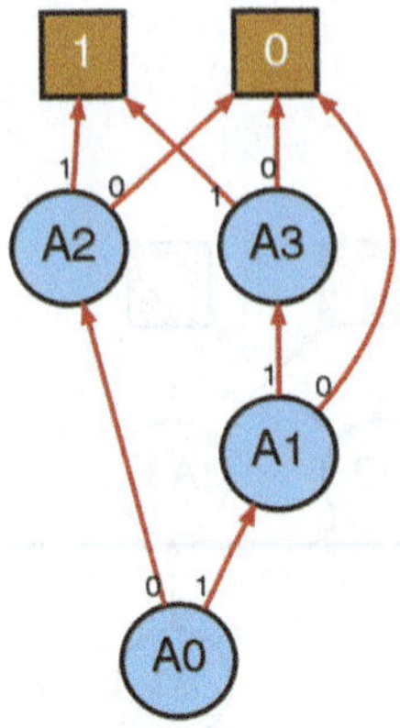

Template AXO210V1

Description: 2-Input AND and 2-Input XOR into 2-Input OR Gate

Function: $Y = A \cdot B + A \oplus C$

Variants: AXO210V2, AXO210V3

Inputs	Output
$A \Rightarrow A_0$ $B \Rightarrow \overline{A_1}$ $C \Rightarrow \overline{A_2}$ $C \Rightarrow A_3$	$\overline{Y}$

Template AO21V2

Description: 2-Input AND into 2-Input OR Gate

Function: $Y = A \cdot B + C$

Variants: AO21V1, AO21V3

Inputs	Output
$A \Rightarrow A_0$ $B \Rightarrow \overline{A_3}$ $C \Rightarrow \overline{A_1}$ $C \Rightarrow \overline{A_2}$	$\overline{Y}$

Template AM21V1

Description: 2-to-1 Multiplexer with one AND2 Data Input

Function: $Y = \overline{S} \cdot D_{0A} \cdot D_{0B} + S \cdot D_1$

Variants: AM21V1, AM21V2, AM21V3, AM21V4, AM21V5

Inputs	Output
$D_{0A} \Rightarrow A_1$ $D_{0B} \Rightarrow A_3$ $D_1 \Rightarrow A_2$ $S \Rightarrow \overline{A_0}$	Y

Footprint F8

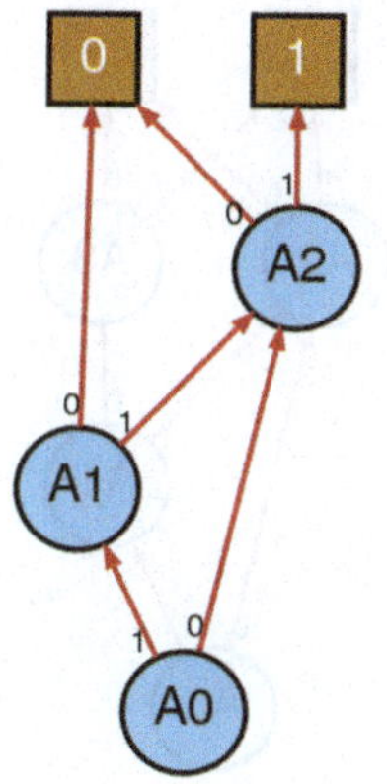

Template AO21V3

Description: 2-Input AND into 2-Input OR Gate

Function: $Y = A \cdot B + C$

Variants: AO21V1, AO21V2

Inputs	Output
$A \Rightarrow A_0$ $B \Rightarrow \overline{A_1}$ $C \Rightarrow \overline{A_2}$	$\overline{Y}$

Footprint F9

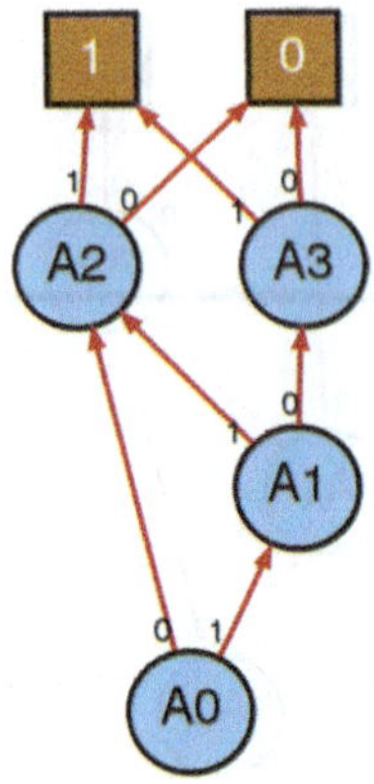

Template AX21V1

Description: 2-Input AND into 2-Input XOR Gate

Function: $Y = A \cdot B \oplus C$

Variants: AX21V2, AX21V3

Inputs	Output
$A \Rightarrow A_0$ $B \Rightarrow \overline{A_1}$ $C \Rightarrow A_2$ $C \Rightarrow \overline{A_3}$	Y

Template AM22V1

Description: 2-to-1 Multiplexer with AND2 Select Input

Function: $Y = SA \cdot SB \cdot D_1 + \overline{SA \cdot SB} \cdot D_0$

Variants: AM22V2, AM22V3

Inputs	Output
$D_0 \Rightarrow A_2$ $D_1 \Rightarrow A_3$ $SA \Rightarrow A_0$ $SB \Rightarrow \overline{A_1}$	Y

Footprint F10

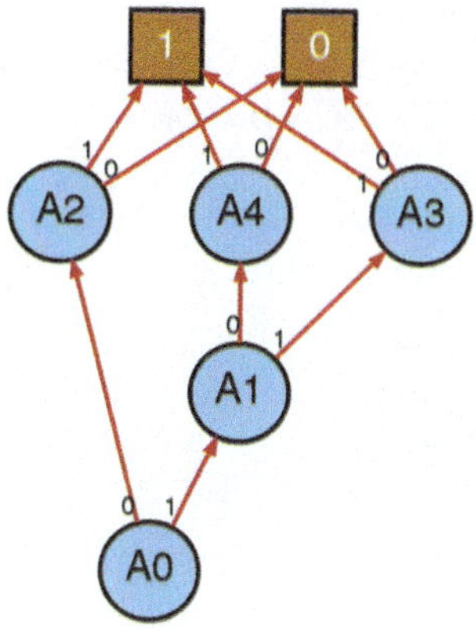

Template AX21V2

Description: 2-Input AND into 2-Input XOR Gate

Function: $Y = A \cdot B \oplus C$

Variants: AX21V1, AX21V3

Inputs	Output
$A \Rightarrow A_0$ $B \Rightarrow A_3$ $B \Rightarrow \overline{A_4}$ $C \Rightarrow A_1$ $C \Rightarrow \overline{A_2}$	$\overline{Y}$

Template XM21V1

Description: 2-to-1 Multiplexer with one XOR2 Data Input

Function: $Y = \overline{S} \cdot (D_{0A} \oplus D_{0B}) + S \cdot D_1$

Variants: XM21V1, XM21V2, XM21V3

Inputs	Output
$D_{0A} \Rightarrow \overline{A_1}$ $D_{0B} \Rightarrow A_3$ $D_{0B} \Rightarrow \overline{A_4}$ $D_1 \Rightarrow A_2$ $S \Rightarrow \overline{A_0}$	Y

Template T4_4

Description:

Function: $Y = A \cdot \overline{B} \cdot \overline{C} + \overline{A} \cdot C + A \cdot \overline{C} \cdot D + A \cdot B \cdot D$

Variants: T4_4, T4_41, T4_32, T4_37

Inputs	Output
$A \Rightarrow A_0$ $B \Rightarrow A_1$ $C \Rightarrow A_2$ $C \Rightarrow \overline{A_4}$ $D \Rightarrow A_3$	Y

Template T4_6

Description:

Function: $Y = \overline{A} \cdot \overline{B} + A \cdot B \cdot C + \overline{B} \cdot D$

Variants: T4_10, T4_17

Inputs	Output
$A \Rightarrow A_0$ $B \Rightarrow A_1$ $B \Rightarrow \overline{A_2}$ $C \Rightarrow A_3$ $D \Rightarrow A_4$	Y

Template MUX3DV1

Description: 3-to-1 Multiplexer (Data overrange output)

Function: $Y = S_1 \cdot S_0 \cdot D_3 + S_1 \cdot \overline{S_0} \cdot D_3 + \overline{S_1} \cdot S_0 \cdot D_1 + \overline{S_1} \cdot \overline{S_0} \cdot D_0$

Variants: MUX3DV2, MUX3DV3, MUX3DV4

Inputs	Output
$D_0 \Rightarrow A_3$ $D_1 \Rightarrow A_4$ $D_3 \Rightarrow A_2$ $S_0 \Rightarrow \overline{A_1}$ $S_1 \Rightarrow \overline{A_0}$	Y

Template MUX3DV2

Description: 3-to-1 Multiplexer (Data overrange output)

Function: $Y = S_1 \cdot S_0 \cdot D_3 + S_1 \cdot \overline{S_0} \cdot D_3 + \overline{S_1} \cdot S_0 \cdot D_1 + \overline{S_1} \cdot \overline{S_0} \cdot D_0$

Variants: MUX3DV1, MUX3DV3, MUX3DV4

Inputs	Output
$D_0 \Rightarrow A_3$ $D_1 \Rightarrow A_4$ $D_3 \Rightarrow A_2$ $S_0 \Rightarrow \overline{A_1}$ $S_1 \Rightarrow \overline{A_0}$	Y

Footprint F11

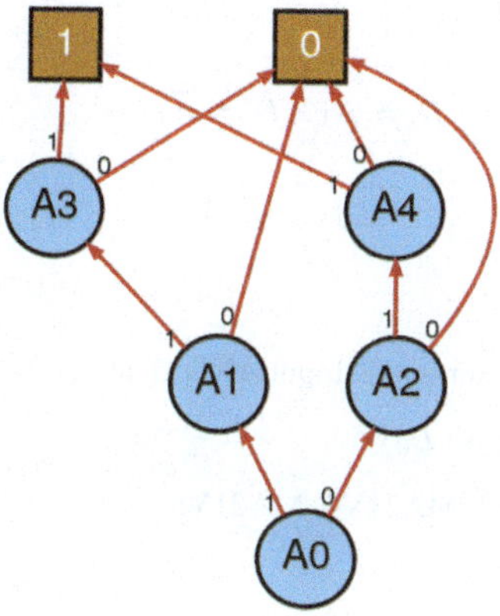

Template CMP3

Description: 3-Input Comparator

Function: $Y = A \cdot B \cdot C + \overline{A} \cdot \overline{B} \cdot \overline{C}$

Variants: None

Inputs	Output
$A \Rightarrow A_0 \quad B \Rightarrow A_1 \quad B \Rightarrow \overline{A_2} \quad C \Rightarrow A_3 \quad C \Rightarrow \overline{A_4}$	Y

Template XA21V2

Description: 2-Input Exclusive-OR into 2-Input AND Gate

Function: $Y = (A \oplus B) \cdot C$

Variants: XA21V1, XA21V3

Inputs	Output
$A \Rightarrow \overline{A_0} \quad B \Rightarrow A_3 \quad B \Rightarrow \overline{A_4} \quad C \Rightarrow A_1 \quad C \Rightarrow A_2$	Y

Template T4_7

Description:

Function: $Y = \overline{A} \cdot C \cdot \overline{D} + A \cdot B \cdot D$

Variants: T4_16

Inputs	Output
$A \Rightarrow A_0$ $B \Rightarrow A_1$ $C \Rightarrow A_2$ $D \Rightarrow A_3$ $D \Rightarrow \overline{A_4}$	Y

Template MA21V2

Description: 2-to-1 Multiplexer into 2-Input AND Gate

Function: $Y = A \cdot (S \cdot D_1 + \overline{S} \cdot D_0)$

Variants: MA21V1, MA21V3, MA21V4, MA21V5

Inputs	Output
$A \Rightarrow A_2$ $A \Rightarrow A_3$ $D_0 \Rightarrow A_1$ $D_1 \Rightarrow A_4$ $S \Rightarrow \overline{A_0}$	Y

Template AM21V2

Description: 2-to-1 Multiplexer with one AND2 Data Input

Function: $Y = \overline{S} \cdot D_{0A} \cdot D_{0B} + S \cdot D_1$

Variants: AM21V1, AM21V3, AM21V4, AM21V5

Inputs	Output
$D_{0A} \Rightarrow A_2$ $D_{0B} \Rightarrow A_4$ $D_1 \Rightarrow A_1$ $D_1 \Rightarrow A_3$ $S \Rightarrow A_0$	Y

Template AAM220

Description: 2-to-1 Multiplexer with AND2 Data Inputs

Function: $Y = S \cdot D_{1A} \cdot D_{1B} + \overline{S} \cdot D_{0A} \cdot D_{0B}$

Variants: None

Inputs	Output
$D_{0A} \Rightarrow A_1$ $D_{0B} \Rightarrow A_3$ $D_{1A} \Rightarrow A_2$ $D_{1B} \Rightarrow A_4$ $S \Rightarrow \overline{A_0}$	Y

Footprint F12

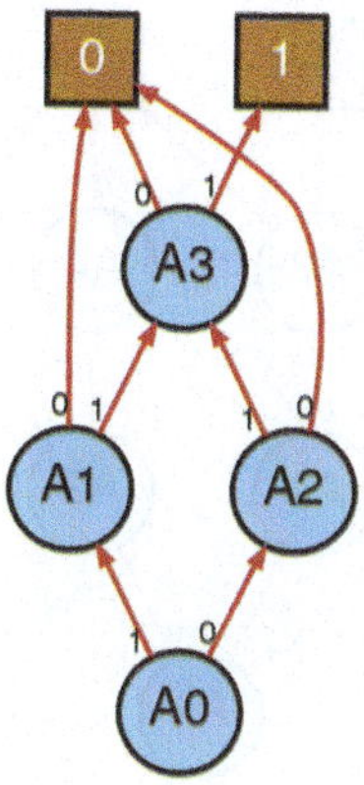

Template XA21V3

Description: 2-Input Exclusive-OR into 2-Input AND Gate

Function: $Y = (A \oplus B) \cdot C$

Variants: XA21V1, XA21V2

Inputs	Output
$A \Rightarrow \overline{A_0}$ $B \Rightarrow A_1$ $B \Rightarrow \overline{A_2}$ $C \Rightarrow A_3$	Y

Template MA21V3

Description: 2-to-1 Multiplexer into 2-Input AND Gate

Function: $Y = A \cdot (S \cdot D_1 + \overline{S} \cdot D_0)$

Variants: MA21V1, MA21V2, MA21V4, MA21V5

Inputs	Output
$A \Rightarrow A_3$ $D_0 \Rightarrow A_1$ $D_1 \Rightarrow A_2$ $S \Rightarrow \overline{A_0}$	Y

Footprint F13

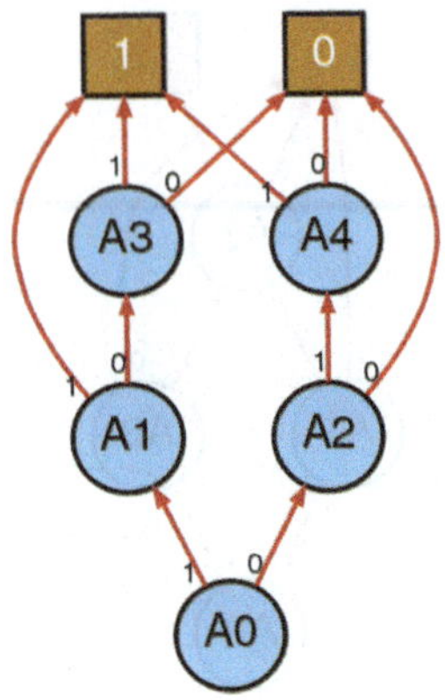

Template MUX2V2

Description: 2-to-1 Multiplexer

Function: $Y = S \cdot D_1 + \overline{S} \cdot D_0$

Variants: MUX2V1, MUX2V3

Inputs	Output
$D_0 \Rightarrow \overline{A_0}$ $D_1 \Rightarrow \overline{A_1}$ $D_1 \Rightarrow \overline{A_2}$ $S \Rightarrow \overline{A_3}$ $S \Rightarrow A_4$	$\overline{Y}$

Template AX21V3

Description: 2-Input AND into 2-Input XOR Gate

Function: $Y = A \cdot B \oplus C$

Variants: AX21V1, AX21V2

Inputs	Output
$A \Rightarrow \overline{A_1}$ $A \Rightarrow A_2$ $B \Rightarrow \overline{A_3}$ $B \Rightarrow A_4$ $C \Rightarrow A_0$	Y

Template AXO210V2

Description: 2-Input AND and 2-Input XOR into 2-Input OR Gate

Function: $Y = A \cdot B + A \oplus C$

Variants: AXO210V1, AXO210V3

Inputs	Output
$A \Rightarrow A_0$ $B \Rightarrow A_3$ $C \Rightarrow \overline{A_1}$ $C \Rightarrow A_2$ $C \Rightarrow A_4$	Y

Template T4_10

Description:

Function: $Y = A \cdot \overline{B} + \overline{A} \cdot C \cdot \overline{D} + A \cdot D$

Variants: T4_6, T4_17

Inputs	Output
$A \Rightarrow A_0$ $B \Rightarrow \overline{A_1}$ $C \Rightarrow A_2$ $D \Rightarrow A_3$ $D \Rightarrow \overline{A_4}$	Y

Template T4_12

Description:

Function: $Y = A \cdot \overline{B} + A \cdot C + C \cdot D$

Variants: T4_11, T4_22

Inputs	Output
$A \Rightarrow A_0$ $B \Rightarrow \overline{A_1}$ $C \Rightarrow A_2$ $C \Rightarrow A_3$ $D \Rightarrow A_4$	Y

Template AOM220

Description: 2-to-1 Multiplexer with one AND2 and one OR2 Data Inputs

Function: $Y = S \cdot (D_{1A} + D_{1B}) + \overline{S} \cdot D_{0A} \cdot D_{0B}$

Variants: None

Inputs	Output
$D_{0A} \Rightarrow \overline{A_1}$ $D_{0B} \Rightarrow \overline{A_3}$ $D_{1A} \Rightarrow \overline{A_2}$ $D_{1B} \Rightarrow \overline{A_4}$ $S \Rightarrow \overline{A_0}$	$\overline{Y}$

Footprint F14

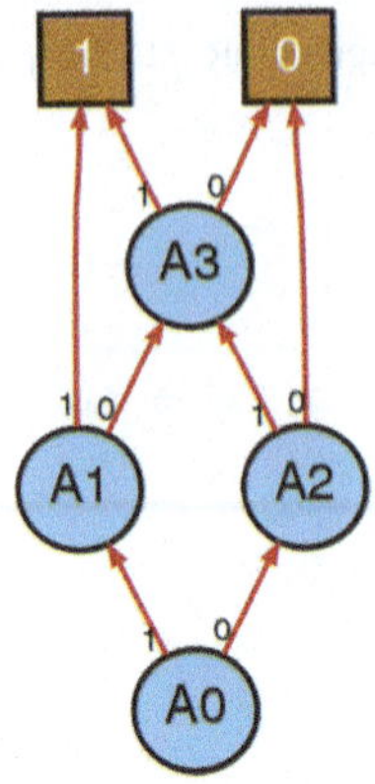

Template MAJ32

Description: 2-out-of-3 Majority Gate

Function: $Y = A \cdot B + A \cdot C + B \cdot C$

Variants: None

Inputs	Output
$A \Rightarrow A_0 \quad B \Rightarrow A_1 \quad B \Rightarrow A_2 \quad C \Rightarrow A_3$	Y

Template MUX2V3

Description: 2-to-1 Multiplexer

Function: $Y = S \cdot D_1 + \overline{S} \cdot D_0$

Variants: MUX2V1, MUX2V2

Inputs	Output
$D_0 \Rightarrow A_0 \quad D_1 \Rightarrow A_3 \quad S \Rightarrow \overline{A_1} \quad S \Rightarrow A_2$	Y

Template T4_11

Description:

Function: $Y = A \cdot \overline{B} + A \cdot D + C \cdot D$

Variants: T4_12, T4_22

Inputs	Output
$A \Rightarrow A_0 \quad B \Rightarrow \overline{A_1} \quad C \Rightarrow A_2 \quad D \Rightarrow A_3$	Y

Footprint F15

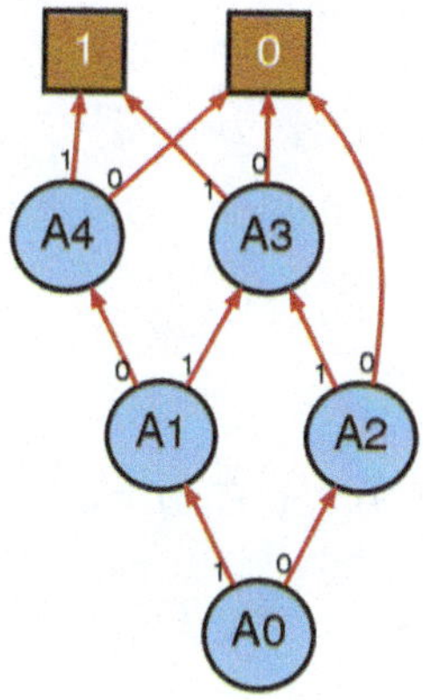

Template T3_18

Description:

Function: $Y = A \cdot \overline{B} \cdot \overline{C} + \overline{A} \cdot \overline{B} \cdot C + A \cdot B \cdot C$

Variants: None

Inputs	Output
$A \Rightarrow A_0$ $B \Rightarrow A_1$ $B \Rightarrow \overline{A_2}$ $C \Rightarrow A_3$ $C \Rightarrow \overline{A_4}$	Y

Template AXO210V3

Description: 2-Input AND and 2-Input XOR into 2-Input OR Gate

Function: $Y = A \cdot B + A \oplus C$

Variants: AXO210V1, AXO210V2

Inputs	Output
$A \Rightarrow \overline{A_1}$ $A \Rightarrow \overline{A_2}$ $B \Rightarrow \overline{A_0}$ $C \Rightarrow \overline{A_3}$ $C \Rightarrow A_4$	$\overline{Y}$

Template T4_13

Description:

Function: $Y = A \cdot \overline{B} \cdot \overline{D} + A \cdot B \cdot D + \overline{A} \cdot C \cdot D$

Variants: T4_18, T4_31

Inputs	Output
$A \Rightarrow A_0$ $B \Rightarrow A_1$ $C \Rightarrow A_2$ $D \Rightarrow A_3$ $D \Rightarrow \overline{A_4}$	Y

Template T4_20

Description:

Function: $Y = A \cdot B \cdot \overline{C} + \overline{A} \cdot C \cdot D + A \cdot \overline{C} \cdot D + A \cdot \overline{B} \cdot D$

Variants: T4_15, T4_43

Inputs	Output
$A \Rightarrow A_0$ $B \Rightarrow \overline{A_1}$ $C \Rightarrow A_2$ $C \Rightarrow \overline{A_4}$ $D \Rightarrow A_3$	Y

Template T4_21

Description:

Function: $Y = A \cdot B \cdot C + A \cdot \overline{B} \cdot D + C \cdot D$

Variants: T4_48

Inputs	Output
$A \Rightarrow A_0$ $B \Rightarrow \overline{A_1}$ $C \Rightarrow A_2$ $C \Rightarrow A_4$ $D \Rightarrow A_3$	Y

Template T4_22

Description:

Function: $Y = A \cdot \overline{B} + \overline{B} \cdot C + A \cdot D$

Variants: T4_12, T4_11

Inputs	Output
$A \Rightarrow A_0$ $B \Rightarrow \overline{A_1}$ $B \Rightarrow \overline{A_3}$ $C \Rightarrow A_2$ $D \Rightarrow A_4$	Y

Template MA21V4

Description: 2-to-1 Multiplexer into 2-Input AND Gate

Function: $Y = A \cdot (S \cdot D_1 + \overline{S} \cdot D_0)$

Variants: MA21V1, MA21V2, MA21V3, MA21V5

Inputs	Output
$A \Rightarrow \overline{A_1}$ $A \Rightarrow A_3$ $D_0 \Rightarrow A_2$ $D_1 \Rightarrow A_4$ $S \Rightarrow A_0$	Y

Template AM21V3

Description: 2-to-1 Multiplexer with one AND2 Data Input

Function: $Y = \overline{S} \cdot D_{0A} \cdot D_{0B} + S \cdot D_1$

Variants: AM21V1, AM21V2, AM21V4, AM21V5

Inputs	Output
$D_{0A} \Rightarrow A_0$ $D_{0B} \Rightarrow A_4$ $D_1 \Rightarrow A_3$ $S \Rightarrow A_1$ $S \Rightarrow A_2$	Y

Template T4_25

Description:

Function: $Y = A \cdot B \cdot C + A \cdot \overline{B} \cdot D + \overline{A} \cdot B \cdot D$

Variants: T4_28, T4_50

Inputs	Output
$A \Rightarrow A_0$ $B \Rightarrow \overline{A_1}$ $B \Rightarrow A_2$ $C \Rightarrow A_4$ $D \Rightarrow A_3$	Y

Template T5_5

Description:

Function: $Y = A \cdot B \cdot D + A \cdot \overline{B} \cdot E + \overline{A} \cdot C \cdot E$

Variants: T5_7, T5_20

Inputs	Output
$A \Rightarrow A_0$ $B \Rightarrow \overline{A_1}$ $C \Rightarrow A_2$ $D \Rightarrow A_4$ $E \Rightarrow A_3$	Y

Footprint F16

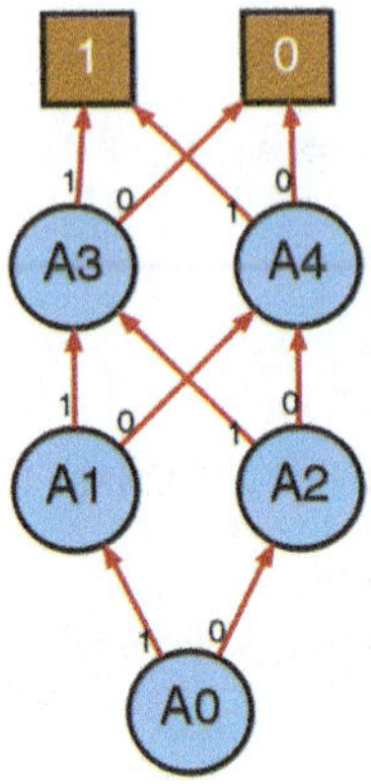

Template XOR3

Description: 3-Input Exclusive-OR Gate

Function: $Y = (A \oplus B) \oplus C$

Variants: None

Inputs	Output
$A \Rightarrow A_0$ $B \Rightarrow A_1$ $B \Rightarrow \overline{A_2}$ $C \Rightarrow A_3$ $C \Rightarrow \overline{A_4}$	Y

Template MX21V1

Description: 2-to-1 Multiplexer into 2-Input XOR Gate

Function: $Y = A \oplus (S \cdot D_1 + \overline{S} \cdot D_0)$

Variants: MX21V2

Inputs	Output
$A \Rightarrow A_3$ $A \Rightarrow \overline{A_4}$ $D_0 \Rightarrow \overline{A_1}$ $D_1 \Rightarrow \overline{A_2}$ $S \Rightarrow \overline{A_0}$	Y

Template T4_48

Description:

Function: $Y = \overline{B} \cdot \overline{C} + A \cdot \overline{B} + A \cdot D + C \cdot D$

Variants: T4_21

Inputs	Output
$A \Rightarrow A_0 \quad B \Rightarrow A_1 \quad B \Rightarrow \overline{A_4} \quad C \Rightarrow A_2 \quad D \Rightarrow A_3$	Y

Template XM22

Description: 2-to-1 Multiplexer with XOR2 Select Input

Function: $Y = (SA \oplus SB) \cdot D_1 + \overline{SA \oplus SB} \cdot D_0$

Variants: None

Inputs	Output
$D_0 \Rightarrow A_3 \quad D_1 \Rightarrow A_4 \quad SA \Rightarrow A_0 \quad SB \Rightarrow A_1 \quad SB \Rightarrow \overline{A_2}$	Y

Template MM22

Description: 2-to-1 Multiplexer with MUX2 Select Input

Function: $Y = (S \cdot S_1 + \overline{S} \cdot S_0) \cdot D_1 + \overline{S \cdot S_1 + \overline{S} \cdot S_0} \cdot D_0$

Variants: None

Inputs	Output
$D_0 \Rightarrow A_3 \quad D_1 \Rightarrow A_4 \quad S \Rightarrow \overline{A_0} \quad S_0 \Rightarrow \overline{A_1} \quad S_1 \Rightarrow \overline{A_2}$	Y

Footprint F17

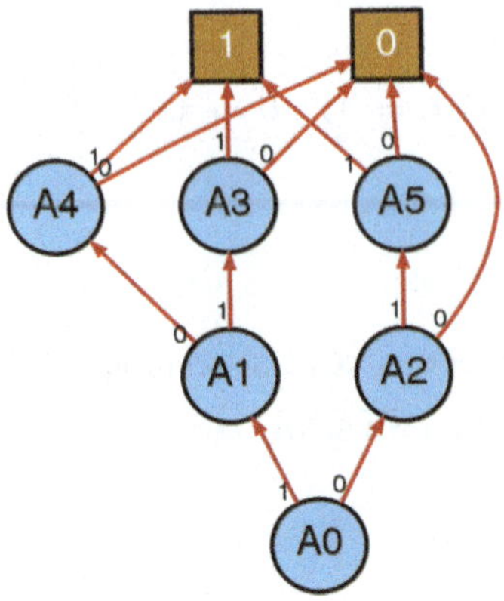

Template T4_14

Description:

Function: $Y = A \cdot B \cdot \overline{C} + \overline{A} \cdot C \cdot \overline{D} + A \cdot \overline{B} \cdot D$

Variants: None

Inputs	Output
$A \Rightarrow A_0$ $B \Rightarrow A_1$ $C \Rightarrow A_2$ $C \Rightarrow \overline{A_3}$ $D \Rightarrow A_4$ $D \Rightarrow \overline{A_5}$	Y

Template T4_15

Description:

Function: $Y = \overline{A} \cdot C \cdot \overline{D} + A \cdot \overline{B} \cdot D + B \cdot C \cdot \overline{D} + A \cdot B \cdot C$

Variants: T4_20, T4_43

Inputs	Output
$A \Rightarrow A_0$ $B \Rightarrow A_1$ $C \Rightarrow A_2$ $C \Rightarrow A_3$ $D \Rightarrow A_4$ $D \Rightarrow \overline{A_5}$	Y

Template T4_16

Description:

Function: $Y = \overline{A} \cdot \overline{B} \cdot C + A \cdot B \cdot D$

Variants: T4_7

Inputs	Output
$A \Rightarrow A_0$ $B \Rightarrow A_1$ $B \Rightarrow A_4$ $B \Rightarrow \overline{A_5}$ $C \Rightarrow A_2$ $D \Rightarrow A_3$	Y

Template T4_17

Description:

Function: $Y = A \cdot \overline{B} + \overline{A} \cdot B \cdot C + A \cdot D$

Variants: T4_6, T4_10

Inputs	Output
$A \Rightarrow A_0$ $B \Rightarrow A_1$ $B \Rightarrow \overline{A_4}$ $B \Rightarrow A_5$ $C \Rightarrow A_2$ $D \Rightarrow A_3$	Y

Template T4_18

Description:

Function: $Y = \overline{A} \cdot \overline{B} \cdot \overline{D} + A \cdot B \cdot C + A \cdot \overline{B} \cdot D$

Variants: T4_13, T4_31

Inputs	Output
$A \Rightarrow A_0$ $B \Rightarrow A_1$ $B \Rightarrow \overline{A_2}$ $C \Rightarrow A_3$ $D \Rightarrow A_4$ $D \Rightarrow \overline{A_5}$	Y

Template T4_19

Description:

Function: $Y = \overline{A} \cdot B \cdot \overline{D} + A \cdot \overline{B} \cdot D + B \cdot C \cdot \overline{D} + A \cdot B \cdot C$

Variants: None

Inputs	Output
$A \Rightarrow A_0$ $B \Rightarrow A_1$ $B \Rightarrow A_2$ $C \Rightarrow A_3$ $D \Rightarrow A_4$ $D \Rightarrow \overline{A_5}$	Y

Template AM21V4

Description: 2-to-1 Multiplexer with one AND2 Data Input

Function: $Y = \overline{S} \cdot D_{0A} \cdot D_{0B} + S \cdot D_1$

Variants: AM21V1, AM21V2, AM21V3, AM21V5

Inputs	Output
$D_{0A} \Rightarrow A_0$ $D_{0B} \Rightarrow A_3$ $D_1 \Rightarrow A_2$ $D_1 \Rightarrow A_4$ $S \Rightarrow \overline{A_1}$ $S \Rightarrow A_5$	Y

Template T4_28

Description:

Function: $Y = A \cdot \overline{B} \cdot C + \overline{A} \cdot B \cdot C + A \cdot B \cdot D$

Variants: T4_25, T4_50

Inputs	Output
$A \Rightarrow A_0$ $B \Rightarrow A_1$ $B \Rightarrow A_5$ $C \Rightarrow A_2$ $C \Rightarrow A_4$ $D \Rightarrow A_3$	Y

Template XM21V2

Description: 2-to-1 Multiplexer with one XOR2 Data Input

Function: $Y = \overline{S} \cdot (D_{0A} \oplus D_{0B}) + S \cdot D_1$

Variants: XM21V1, XM21V3

Inputs	Output
$D_{0A} \Rightarrow A_1$ $D_{0B} \Rightarrow A_3$ $D_{0B} \Rightarrow \overline{A_4}$ $D_1 \Rightarrow \overline{A_2}$ $D_1 \Rightarrow \overline{A_5}$ $S \Rightarrow \overline{A_0}$	$\overline{Y}$

Template MA21V5

Description: 2-to-1 Multiplexer into 2-Input AND Gate

Function: $Y = A \cdot (S \cdot D_1 + \overline{S} \cdot D_0)$

Variants: MA21V1, MA21V2, MA21V3, MA21V4

Inputs	Output
$A \Rightarrow A_0$ $D_0 \Rightarrow A_2$ $D_0 \Rightarrow A_4$ $D_0 \Rightarrow \overline{A_5}$ $D_1 \Rightarrow A_3$ $S \Rightarrow A_1$	Y

Template AM22V2

Description: 2-to-1 Multiplexer with AND2 Select Input

Function: $Y = SA \cdot SB \cdot D_1 + \overline{SA \cdot SB} \cdot D_0$

Variants: AM22V1, AM22V3

Inputs	Output
$D_0 \Rightarrow A_2$ $D_0 \Rightarrow A_4$ $D_1 \Rightarrow A_3$ $SA \Rightarrow A_1$ $SB \Rightarrow A_0$ $SB \Rightarrow \overline{A_5}$	Y

Template T4_41

Description:

Function: $Y = A \cdot \overline{B} + \overline{A} \cdot B \cdot \overline{C} + \overline{B} \cdot C \cdot D + \overline{A} \cdot B \cdot D$

Variants: T4_4, T4_32, T4_37

Inputs	Output
$A \Rightarrow \overline{A_0}$ $B \Rightarrow \overline{A_2}$ $B \Rightarrow A_4$ $B \Rightarrow \overline{A_5}$ $C \Rightarrow A_1$ $D \Rightarrow A_3$	Y

Template T5_4

Description:

Function: $Y = \overline{A} \cdot C \cdot \overline{E} + A \cdot B \cdot D + A \cdot \overline{B} \cdot E$

Variants: None

Inputs	Output
$A \Rightarrow A_0$ $B \Rightarrow A_1$ $C \Rightarrow A_2$ $D \Rightarrow A_3$ $E \Rightarrow A_4$ $E \Rightarrow \overline{A_5}$	Y

Template AXM220

Description: 2-to-1 Multiplexer with one AND2 and one XOR2 Data Inputs

Function: $Y = S \cdot (D_{1A} \oplus D_{1B}) + \overline{S} \cdot D_{0A} \cdot D_{0B}$

Variants: None

Inputs	Output
$D_{0A} \Rightarrow A_2$ $D_{0B} \Rightarrow A_5$ $D_{1A} \Rightarrow \overline{A_1}$ $D_{1B} \Rightarrow A_3$ $D_{1B} \Rightarrow \overline{A_4}$ $S \Rightarrow A_0$	Y

Template T5_7

Description:

Function: $Y = A \cdot \overline{B} \cdot C + A \cdot B \cdot D + \overline{A} \cdot C \cdot E$

Variants: T5_5, T5_20

Inputs	Output
$A \Rightarrow A_0$ $B \Rightarrow A_1$ $C \Rightarrow A_2$ $C \Rightarrow A_4$ $D \Rightarrow A_3$ $E \Rightarrow A_5$	Y

Template MUX3N

Description: 3-to-1 Multiplexer (Zero overrange output)

Function: $Y = S_1 \cdot \overline{S_0} \cdot D_3 + \overline{S_1} \cdot S_0 \cdot D_1 + \overline{S_1} \cdot \overline{S_0} \cdot D_0$

Variants: None

Inputs	Output
$D_0 \Rightarrow A_4$ $D_1 \Rightarrow A_2$ $D_3 \Rightarrow A_3$ $S_0 \Rightarrow \overline{A_0}$ $S_1 \Rightarrow A_1$ $S_1 \Rightarrow \overline{A_5}$	Y

Template MUX3DV3

Description: 3-to-1 Multiplexer (Data overrange output)

Function: $Y = S_1 \cdot S_0 \cdot D_3 + S_1 \cdot \overline{S_0} \cdot D_3 + \overline{S_1} \cdot S_0 \cdot D_1 + \overline{S_1} \cdot \overline{S_0} \cdot D_0$

Variants: MUX3DV1, MUX3DV2, MUX3DV4

Inputs	Output
$D_0 \Rightarrow A_3$ $D_1 \Rightarrow A_4$ $D_3 \Rightarrow A_2$ $D_3 \Rightarrow A_5$ $S_0 \Rightarrow \overline{A_1}$ $S_1 \Rightarrow \overline{A_0}$	Y

Template AMM220

Description: 2-to-1 Multiplexer with one AND2 and one MUX2 Data Inputs

Function: $Y = S \cdot (D_{1S} \cdot D_{11} + \overline{D_{1S}} \cdot D_{10}) + \overline{S} \cdot D_{0A} \cdot D_{0B}$

Variants: None

Inputs	Output
$D_{0A} \Rightarrow A_2$ $D_{0B} \Rightarrow A_5$ $D_{10} \Rightarrow A_3$ $D_{11} \Rightarrow A_4$ $D_{1S} \Rightarrow \overline{A_1}$ $S \Rightarrow A_0$	Y

Footprint F18

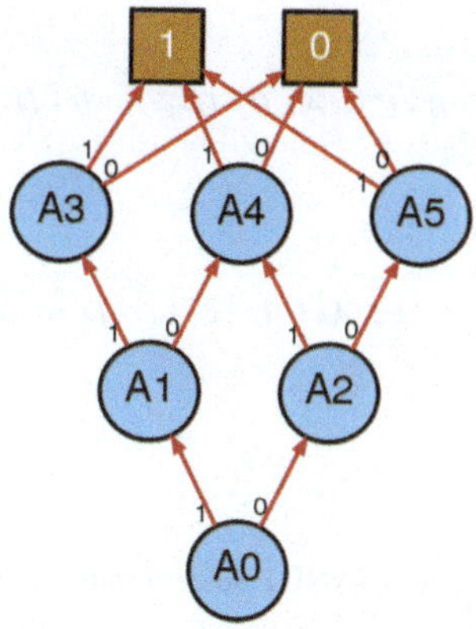

Template T4_30

Description:

Function: $Y = \overline{A} \cdot \overline{C} \cdot \overline{D} + \overline{A} \cdot C \cdot D + A \cdot B \cdot \overline{C} + A \cdot \overline{C} \cdot D + A \cdot \overline{B} \cdot D$

Variants: None

Inputs	Output
$A \Rightarrow A_0$ $B \Rightarrow A_1$ $C \Rightarrow A_2$ $C \Rightarrow A_3$ $D \Rightarrow \overline{A_4}$ $D \Rightarrow A_5$	$\overline{Y}$

Template T4_31

Description:

Function: $Y = \overline{A} \cdot \overline{B} \cdot \overline{C} + \overline{A} \cdot B \cdot C + A \cdot B \cdot D$

Variants: T4_13, T4_18

Inputs	Output
$A \Rightarrow A_0$ $B \Rightarrow A_1$ $B \Rightarrow A_4$ $B \Rightarrow \overline{A_5}$ $C \Rightarrow A_2$ $D \Rightarrow A_3$	Y

Template T4_32

Description:

Function: $Y = \overline{A} \cdot B \cdot \overline{D} + \overline{B} \cdot D + B \cdot C \cdot \overline{D} + A \cdot B \cdot C$

Variants: T4_4, T4_41, T4_37

Inputs	Output
$A \Rightarrow A_0$ $B \Rightarrow A_1$ $B \Rightarrow \overline{A_2}$ $C \Rightarrow A_3$ $D \Rightarrow A_4$ $D \Rightarrow \overline{A_5}$	Y

Template T4_33

Description:

Function: $Y = \overline{A} \cdot \overline{B} \cdot \overline{D} + A \cdot B \cdot C + A \cdot \overline{B} \cdot D + \overline{A} \cdot B \cdot D$

Variants: T4_38

Inputs	Output
$A \Rightarrow A_0$ $B \Rightarrow A_1$ $B \Rightarrow A_2$ $C \Rightarrow A_3$ $D \Rightarrow A_4$ $D \Rightarrow \overline{A_5}$	Y

Template AM22V3

Description: 2-to-1 Multiplexer with AND2 Select Input

Function: $Y = SA \cdot SB \cdot D_1 + \overline{SA \cdot SB} \cdot D_0$

Variants: AM22V1, AM22V2

Inputs	Output
$D_0 \Rightarrow A_4$ $D_0 \Rightarrow \overline{A_5}$ $D_1 \Rightarrow A_3$ $SA \Rightarrow A_1$ $SB \Rightarrow A_0$ $SB \Rightarrow \overline{A_2}$	Y

Template T4_43

Description:

Function: $Y = \overline{A} \cdot \overline{C} \cdot \overline{D} + A \cdot \overline{B} \cdot D + \overline{A} \cdot \overline{B} \cdot C$

Variants: T4_20, T4_15

Inputs	Output
$A \Rightarrow A_0$ $B \Rightarrow \overline{A_1}$ $B \Rightarrow \overline{A_4}$ $C \Rightarrow A_2$ $D \Rightarrow A_3$ $D \Rightarrow \overline{A_5}$	Y

Template T4_44

Description:

Function: $Y = \overline{A} \cdot B \cdot \overline{D} + A \cdot \overline{B} \cdot D + B \cdot C \cdot \overline{D} + \overline{A} \cdot \overline{B} \cdot C + A \cdot B \cdot C$

Variants: None

Inputs	Output
$A \Rightarrow A_0$ $B \Rightarrow \overline{A_1}$ $B \Rightarrow \overline{A_2}$ $C \Rightarrow A_4$ $D \Rightarrow A_3$ $D \Rightarrow \overline{A_5}$	Y

Template XM21V3

Description: 2-to-1 Multiplexer with one XOR2 Data Input

Function: $Y = \overline{S} \cdot (D_{0A} \oplus D_{0B}) + S \cdot D_1$

Variants: XM21V1, XM21V2

Inputs	Output
$D_{0A} \Rightarrow \overline{A_0}$ $D_{0B} \Rightarrow A_3$ $D_{0B} \Rightarrow \overline{A_5}$ $D_1 \Rightarrow A_4$ $S \Rightarrow \overline{A_1}$ $S \Rightarrow A_2$	Y

Template T4_50

Description:

Function: $Y = \overline{B} \cdot \overline{C} + A \cdot \overline{C} + A \cdot \overline{B} + \overline{A} \cdot C \cdot D$

Variants: T4_25, T4_28

Inputs	Output
$A \Rightarrow A_0$ $B \Rightarrow A_1$ $B \Rightarrow A_4$ $C \Rightarrow \overline{A_2}$ $C \Rightarrow A_3$ $D \Rightarrow \overline{A_5}$	$\overline{Y}$

Template AM21V5

Description: 2-to-1 Multiplexer with one AND2 Data Input

Function: $Y = \overline{S} \cdot D_{0A} \cdot D_{0B} + S \cdot D_1$

Variants: AM21V1, AM21V2, AM21V3, AM21V4

Inputs	Output
$D_{0A} \Rightarrow \overline{A_0}$ $D_{0B} \Rightarrow \overline{A_5}$ $D_1 \Rightarrow A_1$ $D_1 \Rightarrow \overline{A_4}$ $S \Rightarrow A_2$ $S \Rightarrow \overline{A_3}$	$\overline{Y}$

Template T4_52

Description:

Function: $Y = A \cdot B \cdot \overline{C} + \overline{A} \cdot \overline{B} \cdot C + \overline{C} \cdot D + \overline{B} \cdot D$

Variants: None

Inputs	Output
$A \Rightarrow A_0$ $B \Rightarrow A_1$ $B \Rightarrow A_5$ $C \Rightarrow \overline{A_2}$ $C \Rightarrow A_3$ $D \Rightarrow \overline{A_4}$	$\overline{Y}$

Template T4_53

Description:

Function: $Y = A \cdot B \cdot \overline{C} + \overline{A} \cdot B \cdot C + \overline{C} \cdot D + A \cdot \overline{B} \cdot D$

Variants: None

Inputs	Output
$A \Rightarrow A_0$ $B \Rightarrow A_1$ $B \Rightarrow \overline{A_5}$ $C \Rightarrow \overline{A_2}$ $C \Rightarrow A_3$ $D \Rightarrow \overline{A_4}$	$\overline{Y}$

Template T5_9

Description:

Function: $Y = \overline{A} \cdot \overline{C} \cdot \overline{E} + A \cdot B \cdot D + A \cdot \overline{B} \cdot E + \overline{A} \cdot C \cdot E$

Variants: T5_11

Inputs	Output
$A \Rightarrow A_0$ $B \Rightarrow A_1$ $C \Rightarrow A_2$ $D \Rightarrow A_3$ $E \Rightarrow A_4$ $E \Rightarrow \overline{A_5}$	Y

Template T5_15

Description:

Function: $Y = \overline{A} \cdot \overline{C} \cdot \overline{E} + A \cdot \overline{B} \cdot E + B \cdot D \cdot \overline{E} + \overline{A} \cdot C \cdot D + A \cdot B \cdot D$

Variants: None

Inputs	Output
$A \Rightarrow A_0$ $B \Rightarrow \overline{A_1}$ $C \Rightarrow A_2$ $D \Rightarrow A_4$ $E \Rightarrow A_3$ $E \Rightarrow \overline{A_5}$	Y

Template T5_19

Description:

Function: $Y = A \cdot B \cdot \overline{C} + \overline{C} \cdot D + A \cdot \overline{B} \cdot D + \overline{A} \cdot C \cdot E$

Variants: None

Inputs	Output
$A \Rightarrow A_0$ $B \Rightarrow A_1$ $C \Rightarrow \overline{A_2}$ $C \Rightarrow A_3$ $D \Rightarrow \overline{A_4}$ $E \Rightarrow \overline{A_5}$	$\overline{Y}$

Template T5_20

Description:

Function: $Y = \overline{B} \cdot \overline{C} + A \cdot \overline{B} + A \cdot D + \overline{A} \cdot C \cdot E$

Variants: T5_5, T5_7

Inputs	Output
$A \Rightarrow A_0$ $B \Rightarrow A_1$ $B \Rightarrow \overline{A_4}$ $C \Rightarrow \overline{A_2}$ $D \Rightarrow A_3$ $E \Rightarrow A_5$	Y

Template T5_21

Description:

Function: $Y = A \cdot B \cdot C + A \cdot \overline{B} \cdot D + \overline{A} \cdot B \cdot D + \overline{A} \cdot \overline{B} \cdot E$

Variants: None

Inputs	Output
$A \Rightarrow A_0$ $B \Rightarrow A_1$ $B \Rightarrow A_2$ $C \Rightarrow A_3$ $D \Rightarrow A_4$ $E \Rightarrow A_5$	Y

Template MUX3DV4

Description: 3-to-1 Multiplexer (Data overrange output)

Function: $Y = S_1 \cdot S_0 \cdot D_3 + S_1 \cdot \overline{S_0} \cdot D_3 + \overline{S_1} \cdot S_0 \cdot D_1 + \overline{S_1} \cdot \overline{S_0} \cdot D_0$

Variants: MUX3DV1, MUX3DV2, MUX3DV3

Inputs	Output
$D_0 \Rightarrow A_3$ $D_1 \Rightarrow A_5$ $D_3 \Rightarrow A_4$ $S_0 \Rightarrow \overline{A_0}$ $S_1 \Rightarrow \overline{A_1}$ $S_1 \Rightarrow A_2$	Y

Template T6_4

Description:

Function: $Y = A \cdot B \cdot D + \overline{A} \cdot \overline{C} \cdot E + A \cdot \overline{B} \cdot E + \overline{A} \cdot C \cdot F$

Variants: None

Inputs	Output
$A \Rightarrow A_0$ $B \Rightarrow A_1$ $C \Rightarrow \overline{A_2}$ $D \Rightarrow A_3$ $E \Rightarrow A_4$ $F \Rightarrow A_5$	Y

Footprint F19

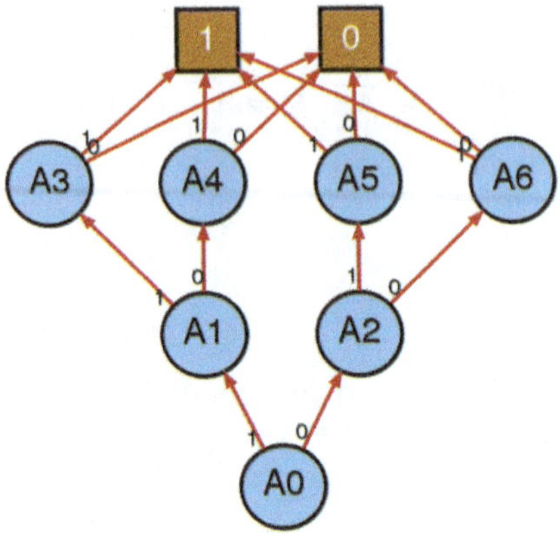

Template MX21V2

Description: 2-to-1 Multiplexer into 2-Input XOR Gate

Function: $Y = A \oplus (S \cdot D_1 + \overline{S} \cdot D_0)$

Variants: MX21V1

Inputs	Output
$A \Rightarrow A_2$ $A \Rightarrow A_3$ $A \Rightarrow \overline{A_4}$ $D_0 \Rightarrow A_1$ $D_1 \Rightarrow A_5$ $D_1 \Rightarrow \overline{A_6}$ $S \Rightarrow \overline{A_0}$	$\overline{Y}$

Template T4_37

Description:

Function: $Y = \overline{B} \cdot \overline{C} + \overline{A} \cdot B \cdot C + A \cdot \overline{C} \cdot D + A \cdot B \cdot D$

Variants: T4_4, T4_41, T4_32

Inputs	Output
$A \Rightarrow A_0$ $B \Rightarrow A_1$ $B \Rightarrow A_5$ $B \Rightarrow \overline{A_6}$ $C \Rightarrow A_2$ $C \Rightarrow \overline{A_4}$ $D \Rightarrow A_3$	Y

Template T4_38

Description:

Function: $Y = A \cdot \overline{B} \cdot \overline{C} + \overline{A} \cdot B \cdot \overline{C} + \overline{A} \cdot \overline{B} \cdot C + A \cdot B \cdot D$

Variants: T4_33

Inputs	Output
$A \Rightarrow A_0$ $B \Rightarrow A_1$ $B \Rightarrow \overline{A_5}$ $B \Rightarrow A_6$ $C \Rightarrow A_2$ $C \Rightarrow \overline{A_4}$ $D \Rightarrow A_3$	Y

Template T4_42

Description:

Function: $Y = \overline{A} \cdot B \cdot \overline{D} + B \cdot C \cdot \overline{D} + \overline{B} \cdot \overline{C} \cdot D + A \cdot B \cdot C + \overline{A} \cdot \overline{C} \cdot \overline{D} + A \cdot \overline{B} \cdot D$

Variants: None

Inputs	Output
$A \Rightarrow A_0$ $B \Rightarrow A_1$ $B \Rightarrow A_2$ $C \Rightarrow A_3$ $C \Rightarrow \overline{A_6}$ $D \Rightarrow A_4$ $D \Rightarrow \overline{A_5}$	Y

Template T4_46

Description:

Function: $Y = A \cdot B \cdot \overline{C} + A \cdot \overline{C} \cdot D + \overline{A} \cdot \overline{B} \cdot C + \overline{A} \cdot \overline{C} \cdot \overline{D} + A \cdot \overline{B} \cdot D$

Variants: None

Inputs	Output
$A \Rightarrow A_0$ $B \Rightarrow A_1$ $B \Rightarrow A_5$ $C \Rightarrow A_2$ $C \Rightarrow A_3$ $D \Rightarrow \overline{A_4}$ $D \Rightarrow A_6$	$\overline{Y}$

Template T4_47

Description:

Function: $Y = \overline{A} \cdot \overline{C} \cdot \overline{D} + A \cdot \overline{B} \cdot D + \overline{A} \cdot B \cdot C + A \cdot B \cdot \overline{C}$

Variants: None

Inputs	Output
$A \Rightarrow A_0$ $B \Rightarrow A_1$ $B \Rightarrow \overline{A_5}$ $C \Rightarrow A_2$ $C \Rightarrow A_3$ $D \Rightarrow \overline{A_4}$ $D \Rightarrow A_6$	$\overline{Y}$

Template XXM220

Description: 2-to-1 Multiplexer with XOR2 Data Inputs

Function: $Y = S \cdot (D_{1A} \oplus D_{1B}) + \overline{S} \cdot (D_{0A} \oplus D_{0B})$

Variants: None

Inputs	Output
$D_{0A} \Rightarrow A_1$ $D_{0B} \Rightarrow A_3$ $D_{0B} \Rightarrow \overline{A_4}$ $D_{1A} \Rightarrow A_2$ $D_{1B} \Rightarrow A_5$ $D_{1B} \Rightarrow \overline{A_6}$ $S \Rightarrow \overline{A_0}$	$\overline{Y}$

Template T5_11

Description:

Function: $Y = \overline{A} \cdot \overline{C} \cdot \overline{E} + A \cdot B \cdot D + A \cdot \overline{B} \cdot \overline{C} + \overline{A} \cdot C \cdot E$

Variants: T5_9

Inputs	Output
$A \Rightarrow A_0$ $B \Rightarrow A_1$ $C \Rightarrow A_2$ $C \Rightarrow \overline{A_4}$ $D \Rightarrow A_3$ $E \Rightarrow A_5$ $E \Rightarrow \overline{A_6}$	Y

Template T5_12

Description:

Function: $Y = \overline{A} \cdot \overline{B} \cdot \overline{C} + \overline{A} \cdot B \cdot C + A \cdot B \cdot D + A \cdot \overline{B} \cdot E$

Variants: None

Inputs	Output
$A \Rightarrow A_0$ $B \Rightarrow A_1$ $B \Rightarrow A_5$ $B \Rightarrow \overline{A_6}$ $C \Rightarrow A_2$ $D \Rightarrow A_3$ $E \Rightarrow A_4$	Y

Template T5_14

Description:

Function: $Y = \overline{A} \cdot \overline{C} \cdot \overline{E} + A \cdot B \cdot D + A \cdot \overline{B} \cdot E + \overline{A} \cdot C \cdot \overline{D}$

Variants: None

Inputs	Output
$A \Rightarrow A_0$ $B \Rightarrow A_1$ $C \Rightarrow A_2$ $D \Rightarrow A_3$ $D \Rightarrow \overline{A_5}$ $E \Rightarrow A_4$ $E \Rightarrow \overline{A_6}$	Y

Template T5_16

Description:

Function: $Y = \overline{A} \cdot \overline{C} \cdot \overline{D} + A \cdot \overline{B} \cdot D + \overline{A} \cdot C \cdot E + A \cdot B \cdot \overline{C}$

Variants: None

Inputs	Output
$A \Rightarrow A_0$ $B \Rightarrow A_1$ $C \Rightarrow A_2$ $C \Rightarrow A_3$ $D \Rightarrow \overline{A_4}$ $D \Rightarrow A_6$ $E \Rightarrow \overline{A_5}$	$\overline{Y}$

Template T5_17

Description:

Function: $Y = \overline{A} \cdot B \cdot \overline{D} + A \cdot \overline{B} \cdot D + B \cdot C \cdot \overline{D} + \overline{A} \cdot \overline{D} \cdot E + \overline{A} \cdot \overline{B} \cdot E + A \cdot B \cdot C$

Variants: None

Inputs	Output
$A \Rightarrow A_0$ $B \Rightarrow A_1$ $B \Rightarrow A_2$ $C \Rightarrow A_3$ $D \Rightarrow A_4$ $D \Rightarrow \overline{A_5}$ $E \Rightarrow A_6$	Y

Template T5_24

Description:

Function: $Y = \overline{B} \cdot \overline{C} + A \cdot B \cdot D + \overline{A} \cdot C \cdot E$

Variants: None

Inputs	Output
$A \Rightarrow A_0$ $B \Rightarrow A_1$ $B \Rightarrow \overline{A_6}$ $C \Rightarrow A_2$ $C \Rightarrow \overline{A_4}$ $D \Rightarrow A_3$ $E \Rightarrow A_5$	Y

Template T5_25

Description:

Function: $Y = A \cdot \overline{B} \cdot \overline{C} + \overline{A} \cdot B \cdot \overline{C} + A \cdot B \cdot D + \overline{A} \cdot C \cdot E$

Variants: None

Inputs	Output
$A \Rightarrow A_0$ $B \Rightarrow A_1$ $B \Rightarrow A_6$ $C \Rightarrow A_2$ $C \Rightarrow \overline{A_4}$ $D \Rightarrow A_3$ $E \Rightarrow A_5$	Y

Template XMM220

Description: 2-to-1 Multiplexer with one XOR2 and one MUX2 Data Inputs

Function: $Y = S \cdot (D_{1S} \cdot D_{11} + \overline{D_{1S}} \cdot D_{10}) + \overline{S} \cdot (D_{0A} \oplus D_{0B})$

Variants: None

Inputs	Output
$D_{0A} \Rightarrow \overline{A_2}$ $D_{0B} \Rightarrow A_5$ $D_{0B} \Rightarrow \overline{A_6}$ $D_{10} \Rightarrow A_3$ $D_{11} \Rightarrow A_4$ $D_{1S} \Rightarrow \overline{A_1}$ $S \Rightarrow A_0$	Y

Template T6_3

Description:

Function: $Y = \overline{A} \cdot \overline{C} \cdot \overline{E} + A \cdot B \cdot D + A \cdot \overline{B} \cdot E + \overline{A} \cdot C \cdot F$

Variants: None

Inputs	Output
$A \Rightarrow A_0$ $B \Rightarrow A_1$ $C \Rightarrow A_2$ $D \Rightarrow A_3$ $E \Rightarrow A_4$ $E \Rightarrow \overline{A_6}$ $F \Rightarrow A_5$	Y

Template T6_5

Description:

Function: $Y = A \cdot \overline{B} \cdot \overline{C} + A \cdot B \cdot D + \overline{A} \cdot C \cdot E + \overline{A} \cdot \overline{C} \cdot F$

Variants: None

Inputs	Output
$A \Rightarrow A_0$ $B \Rightarrow A_1$ $C \Rightarrow A_2$ $C \Rightarrow \overline{A_4}$ $D \Rightarrow A_3$ $E \Rightarrow A_5$ $F \Rightarrow A_6$	Y

Template MUX4

Description: 4-to-1 Multiplexer

Function: $Y = S_1 \cdot (S_0 \cdot D_3 + \overline{S_0} \cdot D_2) + \overline{S_1} \cdot (S_0 \cdot D_1 + \overline{S_0} \cdot D_0)$

Variants: None

Inputs	Output
$D_0 \Rightarrow A_4$ $D_1 \Rightarrow A_3$ $D_2 \Rightarrow A_6$ $D_3 \Rightarrow A_5$ $S_0 \Rightarrow A_1$ $S_0 \Rightarrow A_2$ $S_1 \Rightarrow \overline{A_0}$	Y

Template MMM220

Description: 2-to-1 Multiplexer with MUX2 Data Inputs

Function: $Y = S \cdot (D_{1S} \cdot D_{11} + \overline{D_{1S}} \cdot D_{10}) + \overline{S} \cdot (D_{0S} \cdot D_{01} + \overline{D_{0S}} \cdot D_{00})$

Variants: None

Inputs	Output
$D_{00} \Rightarrow A_3$ $D_{01} \Rightarrow A_4$ $D_{0S} \Rightarrow \overline{A_1}$ $D_{10} \Rightarrow A_5$ $D_{11} \Rightarrow A_6$ $D_{1S} \Rightarrow \overline{A_2}$ $S \Rightarrow \overline{A_0}$	Y

Further Reading

1. M.H. Anis, M.I. Elmasry, Power reduction via an MTCMOS implementation of MOS current mode logic, in *15th Annual IEEE International ASIC/SOC Conference, Sep 2002* (2002), pp. 193–197
2. M.H. Anis, M.I. Elmasry, Self-timed mos current mode logic for digital applications, in *IEEE International Symposium on Circuits and Systems (ISCAS), 2002*, vol. 5 (2002), pp. V-113–V-116
3. S. Badel, I. Hatirnaz, Y. Leblebici, A semi-automated design of a MOS current-mode logic standard cell library from generic components, in *IEEE Conference on Ph.D. Research in Microelectronics and Electronics (PRIME)* (2005)
4. S. Badel, I. Hatirnaz, E.J. Brauer, Y. Leblebici, Implementation of structured ASIC fabric using via-programmable differential MCML cells, in *IFIP International Conference on Very Large Scale Integration (VLSI-SoC)* (2006)
5. S. Badel, E. Güleyüpoğlu, Ö. İnać, A.P. Martinez, P. Vietti, F.K. Gürkaynak, Y. Leblebici, A generic standard cell design methodology for differential circuit styles, in *Design Automation and Test in Europe (DATE)* (2008)
6. E.J. Brauer, S. Badel, I. Hatirnaz, L. Leblebici, Via-programmable expanded universal logic gate in MCML for structured ASIC applications: circuit design, in *IEEE International Symposium on Circuits and Systems (ISCAS)* (2006)
7. V.P. Correia, A.I. Reis, Classifying n-input Boolean functions, in *VII Workshop IBERCHIP* (2001)
8. H. Dang, M. Sawan, Y. Savaria, A novel approach for implementing ultra-high speed flash ADC using MCML circuits, in *IEEE International Symposium on Circuits and Systems (ISCAS)* (2005)
9. H. Hassan, M. Anis, M. Elmasry, Analysis and design of low-power multi-threshold MCML, in *IEEE International SoC Conference* (2004)
10. H. Hassan, M. Anis, M. Elmasry, Low-power multi-threshold MCML: analysis, design, and variability. Microchem. J. **37**, 1097–1104 (2006)
11. I. Hatirnaz, S. Badel, Y. Leblebici, Towards a unified top-down design flow for fully-differential logic blocks with improved speed and noise immunity, in *IEEE Conference on Ph.D. Research in Microelectronics and Electronics (PRIME)* (2005)
12. I. Hatirnaz, S. Badel, E.J. Brauer, Y. Leblebici, Via-programmable structured ASIC fabric based on MCML cells: design flow and implementation, in *IEEE Midwest Symposium on Circuits and Systems (MWSCAS)* (2006)
13. J. Kundan, S.M.R Hasan, Enhanced folded source-coupled logic technique for low-voltage mixed-signal integrated circuits. IEEE Trans. Circuits Syst. **47**, 810–817 (2000)

S. Badel et al., *Design Automation for Differential MOS Current-Mode Logic Circuits*, https://doi.org/10.1007/978-3-319-91307-0

14. T. Kwan, M. Shams, Multi-GHz energy-efficient asynchronous pipelined circuits in MOS current mode logic, in *IEEE International Symposium on Circuits and Systems (ISCAS)* (2004)
15. T.W. Kwan, M. Shams, Design of asynchronous circuit primitives using MOS current-mode logic (MCML), in *International Conference on Microelectronics* (2004)
16. T.W. Kwan, M. Shams, Design of high-performance power-aware asynchronous pipelined circuits in MOS current-mode logic, in *IEEE International Symposium on Asynchronous Circuits and Systems* (2005)
17. F. Macé, F.-X. Standaert, J.-J. Quisquater, J.-D. Legat, A design methodology for secured ICS using dynamic current mode logic, in *Integrated Circuit and System Design* (Springer, Berlin, 2005), pp. 550–560
18. S.R. Maskai, S. Kiaei, D.J. Allstot, Synthesis techniques for CMOS folded source-coupled logic circuits. IEEE J. Solid State Circuits **27**(8), 1157–1167 (1992)
19. A. Mochizuki, T. Hanyu, Low-power multiple-valued current-mode logic using substrate BIAS control. IEICE Trans. Electron. **E87-C**, 582–588 (2004)
20. H.M. Munirul, M. Kameyama, Multiple-valued source-coupled logic VLSI based on adaptive threshold control and its applications, in *IEEE International Symposium on Circuits and Systems (ISCAS)* (2004)
21. J.A. Olstead et al., Noise problems in mixed analog-digital integrated circuits, in *IEEE Custom Integrated Circuits Conference* (1987)
22. R. Pereira-Arroyo, P. Alvarado-Moya, W.H. Krautschneider, Design of a MCML gate library applying multiobjective optimization, in *IEEE Computer Society Annual Symposium on VLSI* (2007)
23. R. Pereira-Arroyo, P. Alvarado-Moya, W.H. Krautschneider, Design of a MCML gate library applying multiobjective optimization, in *Proceedings of the IEEE Computer Society Annual Symposium on VLSI* (2007), pp. 81–85
24. F. Regazzoni, S. Badel, T. Eisenbarth, J. Grobschadl, A. Poschmann, Z. Toprak, M. Macchetti, L. Pozzi, C. Paar, Y. Leblebici, P. Ienne, A simulation-based methodology for evaluating the DPA-resistance of cryptographic functional units with application to CMOS and MCML technologies, in *IEEE International Conference on Embedded Computer Systems: Architectures, Modeling and Simulation* (2007)
25. D. Slepian, On the number of symmetry types of boolean functions of n variables. Can. J. Math. **5**, 185–193 (1953)
26. D.K. Su, M.J. Loinaz, S. Masui, B.A. Wooley, Experimental results and modeling techniques for substrate noise in mixed-signal integrated circuits. IEEE J. Solid State Circuits **28**, 420–430 (1993)
27. K. Tiri, I. Verbauwhede, A VLSI design flow for secure side-channel attack resistant ICs, in *Proceedings of the Design, Automation and Test in Europe Conference and Exhibition (DATE'05)* (2005), pp. 1530–1591
28. K. Tiri, D. Hwang, A. Hodjat, B.C. Lai, S. Yang, P. Schaumont, I. Verbauwhede, Prototype IC with WDDL and differential routing-DPA resistance assessment, in *Cryptographic Hardware and Embedded Systems (CHES), 2005* (2005), pp. 354–365
29. K. Tiri, D. Hwang, A. Hodjat, B. Lai, S. Yang, P. Schaumont, I. Verbauwhede, A side-channel leakage free coprocessor IC in 0.18μm CMOS for embedded AES-based cryptographic and biometric processing, in *Proceedings of the 42nd Annual Conference on Design Automation* (2005), pp. 222–227
30. K. Zhou, Y. Luo, S. Chen, A. Drake, J.F. Macdonald, T. Zhang, Triple-rail MOS current mode logic for high-speed self-timed pipeline applications, in *IEEE International Symposium on Circuits and Systems (ISCAS)* (2006)

Index

S. Badel et al., *Design Automation for Differential MOS Current-Mode Logic Circuits*, https://doi.org/10.1007/978-3-319-91307-0